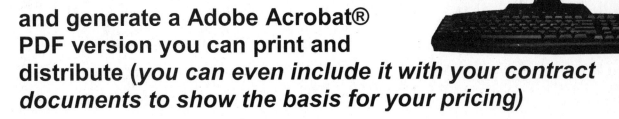

ARCHITECTS CONTRACTORS ENGINEERS

Guide To Construction COSTS

2011 Vol. XLII

Architects Contractors Engineers Guide to Construction Costs 2011, Volume 42.

ISBN 978-1-58855-114-6

EDITOR'S NOTE 2011

This annually published book is designed to give a uniform estimating and cost control system to the General Building Contractor. It contains a complete system to be used with or without computers. It also contains Quick Estimating sections for preliminary conceptual budget estimates by Architects, Engineers and Contractors. Square Foot Estimating is also included for preliminary estimates.

The Metropolitan Area concept is also used and gives the cost modifiers to use for the variations between Metropolitan Areas. This encompasses over 50% of the industry. This book is published annually to be historically accurate with the traditional May-July wage contract settlements and to be a true construction year estimating and cost guide.

The Rate of Inflation in the Construction Industry in 2010 was *2%*. Labor contributed a *3%* increase and materials rose *2%*.

The Wage Rate for Skilled Trades increased an average of *3%* in 2010. Wage rates will probably increase at a *3%* average next year.

The Material Rate increased *2%* in 2010. There were few increases in 2010. The runaway increases of the previous years softened and some materials experienced reductions such as steel, lumber and wood products.

Construction Volume should be down for 2011. Housing, Industrial and Commercial Construction will be down to flat. Highway and Heavy Construction should rise with the stimulus monies finally hitting the market.

The Construction Industry has low to moderate inflation. Some materials should inflate at a slower pace, and some should be watched carefully in 2011.

We are recommending using a *2%* increase in your estimates for work beyond July 1, 2010.

CONTENTS

Metropolitan Cost Index

The costs as presented in this book attempt to represent national averages. Costs, however, vary among regions, states and even between adjacent localities.

In order to more closely approximate the probable costs for specific locations throughout the U.S., this table of Metropolitan Cost Modifiers is provided in the following few pages. These adjustment factors are used to modify costs obtained from this book to help account for regional variations of construction costs and to provide a more accurate estimate for specific areas. The factors are formulated by comparing costs in a specific area to the costs as presented in the Costbook pages. An example of how to use these factors is shown below. Whenever local current costs are known, whether material prices or labor rates, they should be used when more accuracy is required.

Cost Obtained from Costbook Pages   X   **Metroploitan Cost Multiplier Divided by 100**   =   **Adjusted Cost**

For example, a project estimated to cost $125,000 using the Costbook pages can be adjusted to more closely approximate the cost in Los Angeles:

$$\$125,000 \quad X \quad \frac{115}{100} \quad = \quad \$143,750$$

1

State	Metropolitan Area	Multiplier	State	Metropolitan Area	Multiplier
AK	ANCHORAGE	137	FL	OCALA	95
AL	ANNISTON	90		ORLANDO	95
	AUBURN-OPELIKA	90		PANAMA CITY	83
	BIRMINGHAM	88		PENSACOLA	88
	DOTHAN	86		SARASOTA-BRADENTON	89
	GADSDEN	86		TALLAHASSEE	85
	HUNTSVILLE	88		TAMPA-ST. PETERSBURG-CLEARWATER	93
	MOBILE	90		WEST PALM BEACH-BOCA RATON	96
	MONTGOMERY	86	GA	ALBANY	85
	TUSCALOOSA	88		ATHENS	87
AR	FAYETTEVILLE-SPRINGDALE-ROGERS	80		ATLANTA	95
	FORT SMITH	86		AUGUSTA	83
	JONESBORO	85		COLUMBUS	84
	LITTLE ROCK-NORTH LITTLE ROCK	87		MACON	87
	PINE BLUFF	87		SAVANNAH	89
	TEXARKANA	86	HI	HONOLULU	135
AZ	FLAGSTAFF	98	IA	CEDAR RAPIDS	97
	PHOENIX-MESA	98		DAVENPORT	100
	TUCSON	97		DES MOINES	101
	YUMA	99		DUBUQUE	95
CA	BAKERSFIELD	112		IOWA CITY	100
	FRESNO	114		SIOUX CITY	95
	LOS ANGELES-LONG BEACH	115		WATERLOO-CEDAR FALLS	94
	MODESTO	111	ID	BOISE CITY	99
	OAKLAND	120		POCATELLO	97
	ORANGE COUNTY	114	IL	BLOOMINGTON-NORMAL	106
	REDDING	111		CHAMPAIGN-URBANA	105
	RIVERSIDE-SAN BERNARDINO	112		CHICAGO	115
	SACRAMENTO	114		DECATUR	104
	SALINAS	116		KANKAKEE	107
	SAN DIEGO	113		PEORIA-PEKIN	106
	SAN FRANCISCO	125		ROCKFORD	106
	SAN JOSE	122		SPRINGFIELD	104
	SAN LUIS OBISPO	110	IN	BLOOMINGTON	101
	SANTA CRUZ-WATSONVILLE	116		EVANSVILLE	100
	SANTA ROSA	117		FORT WAYNE	100
	STOCKTON-LODI	113		GARY	108
	VALLEJO-FAIRFIELD-NAPA	116		INDIANAPOLIS	104
	VENTURA	112		KOKOMO	100
	SANTA BARBARA	115		LAFAYETTE	100
CO	BOULDER-LONGMONT	96		MUNCIE	100
	COLORADO SPRINGS	101		SOUTH BEND	101
	DENVER	101		TERRE HAUTE	100
	FORT COLLINS-LOVELAND	94	KS	KANSAS CITY	99
	GRAND JUNCTION	96		LAWRENCE	94
	GREELEY	94		TOPEKA	93
	PUEBLO	98		WICHITA	93
CT	BRIDGEPORT	112	KY	LEXINGTON	93
	DANBURY	112		LOUISVILLE	96
	HARTFORD	111		OWENSBORO	94
	NEW HAVEN-MERIDEN	112	LA	ALEXANDRIA	88
	NEW LONDON-NORWICH	109		BATON ROUGE	92
	STAMFORD-NORWALK	115		HOUMA	92
	WATERBURY	111		LAFAYETTE	90
DC	WASHINGTON	104		LAKE CHARLES	92
DE	DOVER	104		MONROE	88
	WILMINGTON-NEWARK	105		NEW ORLEANS	95
FL	DAYTONA BEACH	92		SHREVEPORT-BOSSIER CITY	89
	FORT LAUDERDALE	96	MA	BARNSTABLE-YARMOUTH	115
	FORT MYERS-CAPE CORAL	88		BOSTON	118
	FORT PIERCE-PORT ST. LUCIE	95		BROCKTON	114
	FORT WALTON BEACH	97		FITCHBURG-LEOMINSTER	111
	GAINESVILLE	90		LAWRENCE	114
	JACKSONVILLE	94		LOWELL	109
	LAKELAND-WINTER HAVEN	90		NEW BEDFORD	114
	MELBOURNE-TITUSVILLE-PALM BAY	98		PITTSFIELD	110
	MIAMI	96		SPRINGFIELD	110
	NAPLES	98		WORCESTER	109

State	Metropolitan Area	Multiplier	State	Metropolitan Area	Multiplier
MD	BALTIMORE	98	OH	AKRON	103
	CUMBERLAND	96		CANTON-MASSILLON	101
	HAGERSTOWN	91		CINCINNATI	98
ME	BANGOR	95		CLEVELAND-LORAIN-ELYRIA	106
	LEWISTON-AUBURN	96		COLUMBUS	101
	PORTLAND	97		DAYTON-SPRINGFIELD	99
MI	ANN ARBOR	106		LIMA	101
	DETROIT	111		MANSFIELD	97
	FLINT	105		STEUBENVILLE	104
	GRAND RAPIDS-MUSKEGON-HOLLAND	100		TOLEDO	103
	JACKSON	104		YOUNGSTOWN-WARREN	100
	KALAMAZOO-BATTLE CREEK	96	OK	ENID	88
	LANSING-EAST LANSING	104		LAWTON	89
	SAGINAW-BAY CITY-MIDLAND	102		OKLAHOMA CITY	91
MN	DULUTH	106		TULSA	90
	MINNEAPOLIS-ST. PAUL	111	OR	EUGENE-SPRINGFIELD	106
	ROCHESTER	106		MEDFORD-ASHLAND	104
	ST. CLOUD	108		PORTLAND	108
MO	COLUMBIA	99		SALEM	107
	JOPLIN	95	PA	ALLENTOWN-BETHLEHEM-EASTON	105
	KANSAS CITY	102		ALTOONA	103
	SPRINGFIELD	97		ERIE	103
	ST. JOSEPH	99		HARRISBURG-LEBANON-CARLISLE	101
	ST. LOUIS	99		JOHNSTOWN	104
MS	BILOXI-GULFPORT-PASCAGOULA	87		LANCASTER	101
	JACKSON	86		PHILADELPHIA	114
MT	BILLINGS	98		PITTSBURGH	104
	GREAT FALLS	99		READING	103
	MISSOULA	97		SCRANTON-WILKES-BARRE-HAZLETON	102
NC	ASHEVILLE	82		STATE COLLEGE	98
	CHARLOTTE	83		WILLIAMSPORT	100
	FAYETTEVILLE	84		YORK	102
	GREENSBORO-WINSTON-SALEM-H POINT	83	RI	PROVIDENCE	109
	GREENVILLE	83	SC	AIKEN	91
	HICKORY-MORGANTON-LENOIR	79		CHARLESTON-NORTH CHARLESTON	85
	RALEIGH-DURHAM-CHAPEL HILL	83		COLUMBIA	85
	ROCKY MOUNT	79		FLORENCE	82
	WILMINGTON	84		GREENVILLE-SPARTANBURG-ANDERSON	84
ND	BISMARCK	95		MYRTLE BEACH	90
	FARGO	96	SD	RAPID CITY	89
	GRAND FORKS	94		SIOUX FALLS	90
NE	LINCOLN	92	TN	CHATTANOOGA	88
	OMAHA	96		JACKSON	87
NH	MANCHESTER	100		JOHNSON CITY	84
	NASHUA	98		KNOXVILLE	88
	PORTSMOUTH	92		MEMPHIS	93
NJ	ATLANTIC-CAPE MAY	110		NASHVILLE	92
	BERGEN-PASSAIC	113	TX	ABILENE	86
	JERSEY CITY	112		AMARILLO	87
	MIDDLESEX-SOMERSET-HUNTERDON	108		AUSTIN-SAN MARCOS	88
	MONMOUTH-OCEAN	111		BEAUMONT-PORT ARTHUR	88
	NEWARK	114		BROWNSVILLE-HARLINGEN-SAN BENITO	89
	TRENTON	111		BRYAN-COLLEGE STATION	87
	VINELAND-MILLVILLE-BRIDGETON	109		CORPUS CHRISTI	85
NM	ALBUQUERQUE	96		DALLAS	91
	LAS CRUCES	92		EL PASO	85
	SANTA FE	98		FORT WORTH-ARLINGTON	91
NV	LAS VEGAS	108		GALVESTON-TEXAS CITY	90
	RENO	107		HOUSTON	92
NY	ALBANY-SCHENECTADY-TROY	102		LAREDO	78
	BINGHAMTON	100		LONGVIEW-MARSHALL	83
	BUFFALO-NIAGARA FALLS	108		LUBBOCK	88
	ELMIRA	95		MCALLEN-EDINBURG-MISSION	84
	GLENS FALLS	94		ODESSA-MIDLAND	85
	JAMESTOWN	102		SAN ANGELO	83
	NASSAU-SUFFOLK	115		SAN ANTONIO	89
	NEW YORK	132		TEXARKANA	86
	ROCHESTER	106		TYLER	86
	SYRACUSE	103		VICTORIA	86
	UTICA-ROME	103		WACO	86

State	Metropolitan Area	Multiplier
UT	PROVO-OREM	94
	SALT LAKE CITY-OGDEN	93
VA	CHARLOTTESVILLE	88
	LYNCHBURG	87
	NORFOLK-VA BEACH-NEWPORT NEWS	91
	RICHMOND-PETERSBURG	91
	ROANOKE	85
VT	BURLINGTON	98
WA	BELLINGHAM	112
	BREMERTON	110
	OLYMPIA	108
	RICHLAND-KENNEWICK-PASCO	106
	SEATTLE-BELLEVUE-EVERETT	111
	SPOKANE	108
	TACOMA	111
	YAKIMA	106

State	Metropolitan Area	Multiplier
WI	APPLETON-OSHKOSH-NEENAH	101
	EAU CLAIRE	100
	GREEN BAY	101
	JANESVILLE-BELOIT	101
	KENOSHA	103
	LA CROSSE	99
	MADISON	100
	MILWAUKEE-WAUKESHA	107
	RACINE	103
	WAUSAU	100
WV	CHARLESTON	99
	HUNTINGTON	100
	PARKERSBURG	98
	WHEELING	100
WY	CASPER	92
	CHEYENNE	93

HOW TO USE THIS BOOK

Labor Columns

➢ Units *include* Workers Comp., Unemployment, and FICA on labor (approx. 35%).
➢ Units *do not include* general conditions and equipment (approx. 10%).
➢ Units *do not include* contractors' overhead and profit (approx. 10%).
➢ Units are Government Prevailing wages.

Material Columns

➢ Units *do not include* general conditions and equipment (approx. 10%).
➢ Units *do not include* sales or use taxes (approx. 5%) of material cost.
➢ Units *do not include* contractors' overhead and profit (approx. 10%) and are FOB job site.

Total Columns (Subcontractors)

➢ Units *do not include* general contractors' overhead or profit (approx. 10%).

Quick Estimating Sections - For Preliminary and Conceptual Estimating

➢ Includes all labor, material, general conditions, equipment, taxes.

TABLE OF CONTENTS   PAGE

		UNIT	LABOR	MAT.	TOTAL
01020.10	**ALLOWANCES**				
0090	Overhead				
1000	$20,000 project				
1040	Average	PCT.			20.00
1080	$100,000 project				
1120	Average	PCT.			15.00
1160	$500,000 project				
1180	Average	PCT.			12.00
1220	$1,000,000 project				
1260	Average	PCT.			10.00
1480	Profit				
1500	$20,000 project				
1540	Average	PCT.			15.00
1580	$100,000 project				
1620	Average	PCT.			12.00
1660	$500,000 project				
1700	Average	PCT.			10.00
1740	$1,000,000 project				
1780	Average	PCT.			8.00
2000	Professional fees				
2100	Architectural				
2120	$100,000 project				
2160	Average	PCT.			10.00
2200	$500,000 project				
2240	Average	PCT.			8.00
2280	$1,000,000 project				
2320	Average	PCT.			7.00
4080	Taxes				
5000	Sales tax				
5040	Average	PCT.			5.00
5080	Unemployment				
5120	Average	PCT.			6.50
5200	Social security (FICA)	"			7.85
01050.10	**FIELD STAFF**				
1000	Superintendent				
1020	Minimum	YEAR			78,600
1040	Average	"			114,660
1060	Maximum	"			163,800
1080	Field engineer				
1100	Minimum	YEAR			60,000
1120	Average	"			92,800
1140	Maximum	"			
1160	Foreman				
1180	Minimum	YEAR			43,600
1200	Average	"			70,900
1220	Maximum	"			103,700
1240	Bookkeeper/timekeeper				
1260	Minimum	YEAR			25,100
1280	Average	"			32,800
1300	Maximum	"			54,600
1320	Watchman				
1340	Minimum	YEAR			17,500

		UNIT	LABOR	MAT.	TOTAL
01050.10	**FIELD STAFF, Cont'd...**				
1360	Average	YEAR			21,800
1380	Maximum	"			35,000
01310.10	**SCHEDULING**				
0090	Scheduling for				
1000	$100,000 project				
1040	Average	PCT.			2.00
1080	$500,000 project				
1120	Average	PCT.			1.00
1160	$1,000,000 project				
1200	Average	PCT.			0.80
4000	Scheduling software				
4020	Minimum	EA.			420
4040	Average	"			2,400
4060	Maximum	"			48,050
01410.10	**TESTING**				
1080	Testing concrete, per test				
1100	Minimum	EA.			19.00
1120	Average	"			31.75
1140	Maximum	"			64.00
1160	Soil, per test				
1180	Minimum	EA.			38.25
1200	Average	"			96.00
1220	Maximum	"			250
01500.10	**TEMPORARY FACILITIES**				
1000	Barricades, temporary				
1010	Highway				
1020	Concrete	L.F.	4.06	11.25	15.31
1040	Wood	"	1.62	3.67	5.29
1060	Steel	"	1.35	3.93	5.28
1080	Pedestrian barricades				
1100	Plywood	S.F.	1.35	2.62	3.97
1120	Chain link fence	"	1.35	2.95	4.30
2000	Trailers, general office type, per month				
2020	Minimum	EA.			200
2040	Average	"			330
2060	Maximum	"			660
2080	Crew change trailers, per month				
2100	Minimum	EA.			120
2120	Average	"			130
2140	Maximum	"			200
01505.10	**MOBILIZATION**				
1000	Equipment mobilization				
1020	Bulldozer				
1040	Minimum	EA.			190
1060	Average	"			390
1080	Maximum	"			650
1100	Backhoe/front-end loader				
1120	Minimum	EA.			110
1140	Average	"			190
1160	Maximum	"			430
1180	Crane, crawler type				

		UNIT	LABOR	MAT.	TOTAL

01505.10 MOBILIZATION, Cont'd...

1200	Minimum	EA.			2,050
1220	Average	"			5,030
1240	Maximum	"			10,790
1260	Truck crane				
1280	Minimum	EA.			470
1300	Average	"			720
1320	Maximum	"			1,250
1340	Pile driving rig				
1360	Minimum	EA.			9,310
1380	Average	"			18,610
1400	Maximum	"			33,500

01525.10 CONSTRUCTION AIDS

1000	Scaffolding/staging, rent per month				
1020	Measured by lineal feet of base				
1040	10' high	L.F.			11.75
1060	20' high	"			21.50
1080	30' high	"			30.00
1100	40' high	"			34.50
1120	50' high	"			41.00
1140	Measured by square foot of surface				
1160	Minimum	S.F.			0.52
1180	Average	"			0.90
1200	Maximum	"			1.61
1220	Safety nets, heavy duty, per job				
1240	Minimum	S.F.			0.35
1260	Average	"			0.41
1280	Maximum	"			0.91
1300	Tarpaulins, fabric, per job				
1320	Minimum	S.F.			0.24
1340	Average	"			0.41
1360	Maximum	"			1.06

01525.20 TEMPORARY CONST. SHELTERS

0010	Standard, alum. with fabric, 12'x20'x15' Ht.	S.F.			16.00
0020	12'x20'x20' Ht.	"			18.50

01570.10 SIGNS

0080	Construction signs, temporary				
1000	Signs, 2' x 4'				
1020	Minimum	EA.			34.25
1040	Average	"			82.00
1060	Maximum	"			290
1160	Signs, 8' x 8'				
1180	Minimum	EA.			92.00
1200	Average	"			290
1220	Maximum	"			2,900

01600.10 EQUIPMENT

0080	Air compressor				
1000	60 cfm				
1020	By day	EA.			90.00
1030	By week	"			270
1040	By month	"			820
1200	600 cfm				

		UNIT	LABOR	MAT.	TOTAL
01600.10	**EQUIPMENT, Cont'd...**				
1210	By day	EA.			520
1220	By week	"			1,570
1230	By month	"			4,700
1300	Air tools, per compressor, per day				
1310	Minimum	EA.			34.75
1320	Average	"			43.50
1330	Maximum	"			61.00
1400	Generators, 5 kw				
1410	By day	EA.			87.00
1420	By week	"			260
1430	By month	"			800
1500	Heaters, salamander type, per week				
1510	Minimum	EA.			100
1520	Average	"			150
1530	Maximum	"			310
1600	Pumps, submersible				
1605	50 gpm				
1610	By day	EA.			70.00
1620	By week	"			210
1630	By month	"			620
1675	500 gpm				
1680	By day	EA.			140
1690	By week	"			420
1700	By month	"			1,250
1900	Diaphragm pump, by week				
1920	Minimum	EA.			120
1930	Average	"			210
1940	Maximum	"			430
2000	Pickup truck				
2020	By day	EA.			130
2030	By week	"			380
2040	By month	"			1,180
2080	Dump truck				
2100	6 cy truck				
2120	By day	EA.			350
2130	By week	"			1,040
2140	By month	"			3,130
2300	16 cy truck				
2310	By day	EA.			700
2320	By week	"			2,090
2340	By month	"			6,260
2400	Backhoe, track mounted				
2420	1/2 cy capacity				
2430	By day	EA.			710
2440	By week	"			2,170
2450	By month	"			6,440
2500	1 cy capacity				
2510	By day	EA.			1,130
2520	By week	"			3,390
2530	By month	"			10,180
2600	3 cy capacity				
2620	By day	EA.			3,650

		UNIT	LABOR	MAT.	TOTAL
01600.10	**EQUIPMENT, Cont'd...**				
2640	By week	EA.			10,960
2680	By month	"			32,890
3000	Backhoe/loader, rubber tired				
3005	1/2 cy capacity				
3010	By day	EA.			430
3020	By week	"			1,300
3030	By month	"			3,910
3035	3/4 cy capacity				
3040	By day	EA.			520
3050	By week	"			1,570
3060	By month	"			4,700
3200	Bulldozer				
3205	75 hp				
3210	By day	EA.			610
3220	By week	"			1,830
3230	By month	"			5,480
4000	Cranes, crawler type				
4005	15 ton capacity				
4010	By day	EA.			780
4020	By week	"			2,350
4030	By month	"			7,050
4070	50 ton capacity				
4080	By day	EA.			1,740
4090	By week	"			5,220
4100	By month	"			15,660
4145	Truck mounted, hydraulic				
4150	15 ton capacity				
4160	By day	EA.			740
4170	By week	"			2,220
4180	By month	"			6,400
5380	Loader, rubber tired				
5385	1 cy capacity				
5390	By day	EA.			520
5400	By week	"			1,570
5410	By month	"			4,700
7000	Scraper				
7010	Elevated scraper, not including bulldozer, 12 c.y.				
7020	By day	EA.			1,190
7030	By week	"			3,570
7040	By month	"			10,110
7100	Self-propelled scraper, 14 c.y.				
7110	By day	EA.			2,380
7120	By week	"			7,140
7130	By month	"			20,220
01740.10	**BONDS**				
1000	Performance bonds				
1020	Minimum	PCT.			0.64
1040	Average	"			2.01
1060	Maximum	"			3.18

TABLE OF CONTENTS   PAGE

		UNIT	LABOR	MAT.	TOTAL
02210.10	**SOIL BORING**				
1000	Borings, uncased, stable earth				
1020	2-1/2" dia.	L.F.	28.00		28.00
1040	4" dia.	"	31.75		31.75
1500	Cased, including samples				
1520	2-1/2" dia.	L.F.	37.25		37.25
1540	4" dia.	"	64.00		64.00
02220.10	**COMPLETE BUILDING DEMOLITION**				
0200	Wood frame	C.F.	0.32		0.32
0300	Concrete	"	0.48		0.48
0400	Steel frame	"	0.64		0.64
02220.15	**SELECTIVE BUILDING DEMOLITION**				
1000	Partition removal				
1100	Concrete block partitions				
1120	4" thick	S.F.	2.03		2.03
1140	8" thick	"	2.70		2.70
1160	12" thick	"	3.69		3.69
1200	Brick masonry partitions				
1220	4" thick	S.F.	2.03		2.03
1240	8" thick	"	2.53		2.53
1260	12" thick	"	3.38		3.38
1280	16" thick	"	5.07		5.07
1380	Cast in place concrete partitions				
1400	Unreinforced				
1421	6" thick	S.F.	14.75		14.75
1423	8" thick	"	16.00		16.00
1425	10" thick	"	18.50		18.50
1427	12" thick	"	22.25		22.25
1440	Reinforced				
1441	6" thick	S.F.	17.25		17.25
1443	8" thick	"	22.25		22.25
1445	10" thick	"	24.75		24.75
1447	12" thick	"	29.75		29.75
1500	Terra cotta				
1520	To 6" thick	S.F.	2.03		2.03
1700	Stud partitions				
1720	Metal or wood, with drywall both sides	S.F.	2.03		2.03
1740	Metal studs, both sides, lath and plaster	"	2.70		2.70
2000	Door and frame removal				
2020	Hollow metal in masonry wall				
2030	Single				
2040	2'6"x6'8"	EA.	51.00		51.00
2060	3'x7'	"	68.00		68.00
2070	Double				
2080	3'x7'	EA.	81.00		81.00
2085	4'x8'	"	81.00		81.00
2140	Wood in framed wall				
2150	Single				
2160	2'6"x6'8"	EA.	29.00		29.00
2180	3'x6'8"	"	33.75		33.75
2190	Double				
2200	2'6"x6'8"	EA.	40.50		40.50

		UNIT	LABOR	MAT.	TOTAL
02220.15	**SELECTIVE BUILDING DEMOLITION, Cont'd...**				
2220	3'x6'8"	EA.	45.00		45.00
2240	Remove for re-use				
2260	Hollow metal	EA.	100		100
2280	Wood	"	68.00		68.00
2300	Floor removal				
2340	Brick flooring	S.F.	1.62		1.62
2360	Ceramic or quarry tile	"	0.90		0.90
2380	Terrazzo	"	1.80		1.80
2400	Heavy wood	"	1.08		1.08
2420	Residential wood	"	1.16		1.16
2440	Resilient tile or linoleum	"	0.40		0.40
2500	Ceiling removal				
2520	Acoustical tile ceiling				
2540	Adhesive fastened	S.F.	0.40		0.40
2560	Furred and glued	"	0.33		0.33
2580	Suspended grid	"	0.25		0.25
2600	Drywall ceiling				
2620	Furred and nailed	S.F.	0.45		0.45
2640	Nailed to framing	"	0.40		0.40
2660	Plastered ceiling				
2680	Furred on framing	S.F.	1.01		1.01
2700	Suspended system	"	1.35		1.35
2800	Roofing removal				
2820	Steel frame				
2840	Corrugated metal roofing	S.F.	0.81		0.81
2860	Built-up roof on metal deck	"	1.35		1.35
2900	Wood frame				
2920	Built up roof on wood deck	S.F.	1.24		1.24
2940	Roof shingles	"	0.67		0.67
2960	Roof tiles	"	1.35		1.35
8900	Concrete frame	C.F.	2.70		2.70
8920	Concrete plank	S.F.	2.03		2.03
8940	Built-up roof on concrete	"	1.16		1.16
9200	Cut-outs				
9230	Concrete, elevated slabs, mesh reinforcing				
9240	Under 5 cf	C.F.	40.50		40.50
9260	Over 5 cf	"	33.75		33.75
9270	Bar reinforcing				
9280	Under 5 cf	C.F.	68.00		68.00
9290	Over 5 cf	"	51.00		51.00
9300	Window removal				
9301	Metal windows, trim included				
9302	2'x3'	EA.	40.50		40.50
9304	2'x4'	"	45.00		45.00
9306	2'x6'	"	51.00		51.00
9308	3'x4'	"	51.00		51.00
9310	3'x6'	"	58.00		58.00
9312	3'x8'	"	68.00		68.00
9314	4'x4'	"	68.00		68.00
9315	4'x6'	"	81.00		81.00
9316	4'x8'	"	100		100
9317	Wood windows, trim included				

		UNIT	LABOR	MAT.	TOTAL
02220.15	**SELECTIVE BUILDING DEMOLITION, Cont'd...**				
9318	2'x3'	EA.	22.50		22.50
9319	2'x4'	"	24.00		24.00
9320	2'x6'	"	25.25		25.25
9321	3'x4'	"	27.00		27.00
9322	3'x6'	"	29.00		29.00
9324	3'x8'	"	31.25		31.25
9325	6'x4'	"	33.75		33.75
9326	6'x6'	"	37.00		37.00
9327	6'x8'	"	40.50		40.50
9329	Walls, concrete, bar reinforcing				
9330	Small jobs	C.F.	27.00		27.00
9340	Large jobs	"	22.50		22.50
9360	Brick walls, not including toothing				
9390	4" thick	S.F.	2.03		2.03
9400	8" thick	"	2.53		2.53
9410	12" thick	"	3.38		3.38
9415	16" thick	"	5.07		5.07
9420	Concrete block walls, not including toothing				
9440	4" thick	S.F.	2.25		2.25
9450	6" thick	"	2.38		2.38
9460	8" thick	"	2.53		2.53
9465	10" thick	"	2.90		2.90
9470	12" thick	"	3.38		3.38
9500	Rubbish handling				
9519	Load in dumpster or truck				
9520	Minimum	C.F.	0.90		0.90
9540	Maximum	"	1.35		1.35
9550	For use of elevators, add				
9560	Minimum	C.F.	0.20		0.20
9570	Maximum	"	0.40		0.40
9600	Rubbish hauling				
9640	Hand loaded on trucks, 2 mile trip	C.Y.	34.50		34.50
9660	Machine loaded on trucks, 2 mile trip	"	22.25		22.25
02225.20	**FENCE DEMOLITION**				
0060	Remove fencing				
0080	Chain link, 8' high				
0100	For disposal	L.F.	2.03		2.03
0200	For reuse	"	5.07		5.07
0980	Wood				
1000	4' high	S.F.	1.35		1.35
1960	Masonry				
1980	8" thick				
2000	4' high	S.F.	4.06		4.06
2020	6' high	"	5.07		5.07
02225.50	**SAW CUTTING PAVEMENT**				
0100	Pavement, bituminous				
0110	2" thick	L.F.	1.72		1.72
0120	3" thick	"	2.15		2.15
0200	Concrete pavement, with wire mesh				
0210	4" thick	L.F.	3.31		3.31
0212	5" thick	"	3.59		3.59

		UNIT	LABOR	MAT.	TOTAL
02225.50	**SAW CUTTING PAVEMENT, Cont'd...**				
0300	Plain concrete, unreinforced				
0320	4" thick	L.F.	2.87		2.87
0340	5" thick	"	3.31		3.31
02225.60	**TORCH CUTTING**				
0010	Steel plate, 1/2" thick,	L.F.	0.88	0.49	1.37
0020	1" thick	"	1.77	0.71	2.48
02230.50	**TREE CUTTING & CLEARING**				
0980	Cut trees and clear out stumps				
1000	9" to 12" dia.	EA.	450		450
1400	To 24" dia.	"	560		560
1600	24" dia. and up	"	740		740
02315.10	**BASE COURSE**				
1019	Base course, crushed stone				
1020	3" thick	S.Y.	0.59	3.14	3.73
1030	4" thick	"	0.64	4.21	4.85
1040	6" thick	"	0.70	6.32	7.02
2500	Base course, bank run gravel				
3020	4" deep	S.Y.	0.62	2.56	3.18
3040	6" deep	"	0.68	3.85	4.53
4000	Prepare and roll sub base				
4020	Minimum	S.Y.	0.59		0.59
4030	Average	"	0.74		0.74
4040	Maximum	"	0.99		0.99
02315.20	**BORROW**				
1000	Borrow fill, F.O.B. at pit				
1005	Sand, haul to site, round trip				
1010	10 mile	C.Y.	12.00	16.50	28.50
1020	20 mile	"	19.75	16.50	36.25
1030	30 mile	"	29.75	16.50	46.25
3980	Place borrow fill and compact				
4000	Less than 1 in 4 slope	C.Y.	5.95	16.50	22.45
4100	Greater than 1 in 4 slope	"	7.94	16.50	24.44
02315.30	**BULK EXCAVATION**				
1000	Excavation, by small dozer				
1020	Large areas	C.Y.	1.72		1.72
1040	Small areas	"	2.87		2.87
1060	Trim banks	"	4.31		4.31
1700	Hydraulic excavator				
1720	1 cy capacity				
1740	Light material	C.Y.	3.71		3.71
1760	Medium material	"	4.46		4.46
1780	Wet material	"	5.57		5.57
1790	Blasted rock	"	6.37		6.37
1800	1-1/2 cy capacity				
1820	Light material	C.Y.	1.48		1.48
1840	Medium material	"	1.98		1.98
1860	Wet material	"	2.38		2.38
2000	Wheel mounted front-end loader				
2020	7/8 cy capacity				
2040	Light material	C.Y.	2.97		2.97

DIVISION # 02 SITE CONSTRUCTION

		UNIT	LABOR	MAT.	TOTAL
02315.30	**BULK EXCAVATION, Cont'd...**				
2060	Medium material	C.Y.	3.40		3.40
2080	Wet material	"	3.97		3.97
2100	Blasted rock	"	4.76		4.76
2300	2-1/2 cy capacity				
2320	Light material	C.Y.	1.40		1.40
2340	Medium material	"	1.48		1.48
2360	Wet material	"	1.58		1.58
2380	Blasted rock	"	1.70		1.70
2600	Track mounted front-end loader				
2620	1-1/2 cy capacity				
2640	Light material	C.Y.	1.98		1.98
2660	Medium material	"	2.16		2.16
2680	Wet material	"	2.38		2.38
2700	Blasted rock	"	2.64		2.64
2720	2-3/4 cy capacity				
2740	Light material	C.Y.	1.19		1.19
2760	Medium material	"	1.32		1.32
2780	Wet material	"	1.48		1.48
2790	Blasted rock	"	1.70		1.70
4000	Scraper -500' haul				
4010	Elevated scraper, not including bulldozer, 12 c.y.				
4020	Light material	C.Y.	3.97		3.97
4030	Medium material	"	4.33		4.33
4040	Wet material	"	4.76		4.76
4050	Blasted rock	"	5.29		5.29
4100	Self-propelled scraper, 14 c.y.				
4110	Light material	C.Y.	3.66		3.66
4120	Medium material	"	3.97		3.97
4130	Wet material	"	4.33		4.33
4140	Blasted rock	"	4.76		4.76
4200	1,000' haul				
4210	Elevated scraper, not including bulldozer, 12 c.y.				
4220	Light material	C.Y.	4.76		4.76
4230	Medium material	"	5.29		5.29
4240	Wet material	"	5.95		5.95
4250	Blasted rock	"	6.80		6.80
4300	Self-propelled scraper, 14 c.y.				
4310	Light material	C.Y.	4.33		4.33
4320	Medium material	"	4.76		4.76
4330	Wet material	"	5.29		5.29
4340	Blasted rock	"	5.95		5.95
4400	2,000' haul				
4410	Elevated scraper, not including bulldozer, 12 c.y.				
4420	Light material	C.Y.	5.95		5.95
4430	Medium material	"	6.80		6.80
4440	Wet material	"	7.94		7.94
4450	Blasted rock	"	9.52		9.52
4510	Self-propelled scraper, 14 c.y.				
4520	Light material	C.Y.	5.29		5.29
4530	Medium material	"	5.95		5.95
4540	Wet material	"	6.80		6.80
4550	Blasted rock	"	7.94		7.94

		UNIT	LABOR	MAT.	TOTAL
02315.40	**BUILDING EXCAVATION**				
0090	Structural excavation, unclassified earth				
0100	3/8 cy backhoe	C.Y.	16.00		16.00
0110	3/4 cy backhoe	"	12.00		12.00
0120	1 cy backhoe	"	9.92		9.92
0600	Foundation backfill and compaction by machine	"	23.75		23.75
02315.45	**HAND EXCAVATION**				
0980	Excavation				
1000	To 2' deep				
1020	Normal soil	C.Y.	45.00		45.00
1040	Sand and gravel	"	40.50		40.50
1060	Medium clay	"	51.00		51.00
1080	Heavy clay	"	58.00		58.00
1100	Loose rock	"	68.00		68.00
1200	To 6' deep				
1220	Normal soil	C.Y.	58.00		58.00
1240	Sand and gravel	"	51.00		51.00
1260	Medium clay	"	68.00		68.00
1280	Heavy clay	"	81.00		81.00
1300	Loose rock	"	100		100
2020	Backfilling foundation without compaction, 6" lifts	"	25.25		25.25
2200	Compaction of backfill around structures or in trench				
2220	By hand with air tamper	C.Y.	29.00		29.00
2240	By hand with vibrating plate tamper	"	27.00		27.00
2250	1 ton roller	"	43.00		43.00
5400	Miscellaneous hand labor				
5440	Trim slopes, sides of excavation	S.F.	0.06		0.06
5450	Trim bottom of excavation	"	0.08		0.08
5460	Excavation around obstructions and services	C.Y.	140		140
02315.50	**ROADWAY EXCAVATION**				
0100	Roadway excavation				
0110	1/4 mile haul	C.Y.	2.38		2.38
0120	2 mile haul	"	3.97		3.97
0130	5 mile haul	"	5.95		5.95
3000	Spread base course	"	2.97		2.97
3100	Roll and compact	"	3.97		3.97
02315.60	**TRENCHING**				
0100	Trenching and continuous footing excavation				
0980	By gradall				
1000	1 cy capacity				
1040	Medium soil	C.Y.	3.66		3.66
1080	Loose rock	"	4.33		4.33
1090	Blasted rock	"	4.58		4.58
1095	By hydraulic excavator				
1100	1/2 cy capacity				
1140	Medium soil	C.Y.	4.33		4.33
1180	Loose rock	"	5.29		5.29
1190	Blasted rock	"	5.95		5.95
1200	1 cy capacity				
1240	Medium soil	C.Y.	2.97		2.97
1280	Loose rock	"	3.40		3.40
1300	Blasted rock	"	3.66		3.66

		UNIT	LABOR	MAT.	TOTAL
02315.60	**TRENCHING, Cont'd...**				
1600	2 cy capacity				
1640	Medium soil	C.Y.	2.50		2.50
1680	Loose rock	"	2.80		2.80
1690	Blasted rock	"	2.97		2.97
3000	Hand excavation				
3100	Bulk, wheeled 100'				
3120	Normal soil	C.Y.	45.00		45.00
3140	Sand or gravel	"	40.50		40.50
3160	Medium clay	"	58.00		58.00
3180	Heavy clay	"	81.00		81.00
3200	Loose rock	"	100		100
3300	Trenches, up to 2' deep				
3320	Normal soil	C.Y.	51.00		51.00
3340	Sand or gravel	"	45.00		45.00
3360	Medium clay	"	68.00		68.00
3380	Heavy clay	"	100		100
3390	Loose rock	"	140		140
3400	Trenches, to 6' deep				
3420	Normal soil	C.Y.	58.00		58.00
3440	Sand or gravel	"	51.00		51.00
3460	Medium clay	"	81.00		81.00
3480	Heavy clay	"	140		140
3500	Loose rock	"	200		200
3590	Backfill trenches				
3600	With compaction				
3620	By hand	C.Y.	33.75		33.75
3640	By 60 hp tracked dozer	"	2.15		2.15
02315.70	**UTILITY EXCAVATION**				
2080	Trencher, sandy clay, 8" wide trench				
2100	18" deep	L.F.	1.91		1.91
2200	24" deep	"	2.15		2.15
2300	36" deep	"	2.46		2.46
6080	Trench backfill, 95% compaction				
7000	Tamp by hand	C.Y.	25.25		25.25
7050	Vibratory compaction	"	20.25		20.25
7060	Trench backfilling, with borrow sand, place & compact	"	20.25	16.00	36.25
02315.80	**HAULING MATERIAL**				
0090	Haul material by 10 cy dump truck, round trip distance				
0100	1 mile	C.Y.	4.78		4.78
0110	2 mile	"	5.74		5.74
0120	5 mile	"	7.83		7.83
0130	10 mile	"	8.62		8.62
0140	20 mile	"	9.57		9.57
0150	30 mile	"	11.50		11.50
2000	Site grading, cut & fill, sandy clay, 200' haul, 75 hp dozer	"	3.44		3.44
6000	Spread topsoil by equipment on site	"	3.83		3.83
6980	Site grading (cut and fill to 6") less than 1 acre				
7000	75 hp dozer	C.Y.	5.74		5.74
7600	1.5 cy backhoe/loader	"	8.62		8.62
0010	Blasting mats, up to 1500 c.y.	"	92.00	2.86	94.86
0020	Buried explosives, up to 1500 c.y.	"	27.75	2.86	30.61

		UNIT	LABOR	MAT.	TOTAL
02340.05	**SOIL STABILIZATION**				
0100	Straw bale secured with rebar	L.F.	1.35	7.54	8.89
0120	Filter barrier, 18" high filter fabric	"	4.06	1.82	5.88
0130	Sediment fence, 36" fabric with 6" mesh	"	5.07	4.32	9.39
1000	Soil stabilization with tar paper, burlap, straw and stakes	S.F.	0.05	0.36	0.41
02360.20	**SOIL TREATMENT**				
1100	Soil treatment, termite control pretreatment				
1120	Under slabs	S.F.	0.22	0.38	0.60
1140	By walls	"	0.27	0.38	0.65
02370.40	**RIPRAP**				
0100	Riprap				
0110	Crushed stone blanket, max size 2-1/2"	TON	65.00	35.25	100
0120	Stone, quarry run, 300 lb. stones	"	60.00	44.25	104
0130	400 lb. stones	"	55.00	46.00	101
0140	500 lb. stones	"	52.00	48.00	100
0150	750 lb. stones	"	48.50	49.75	98.25
0160	Dry concrete riprap in bags 3" thick, 80 lb. per bag	BAG	3.22	5.96	9.18
02455.60	**STEEL PILES**				
1000	H-section piles				
1010	8x8				
1020	36 lb/ft				
1021	30' long	L.F.	11.25	14.75	26.00
1022	40' long	"	8.90	14.75	23.65
5000	Tapered friction piles, with fluted steel casing, up to 50'				
5002	With 4000 psi concrete no reinforcing				
5040	12" dia.	L.F.	6.68	18.00	24.68
5060	14" dia.	"	6.85	20.75	27.60
02455.65	**STEEL PIPE PILES**				
1000	Concrete filled, 3000# concrete, up to 40'				
1100	8" dia.	L.F.	9.54	19.25	28.79
1120	10" dia.	"	9.89	23.75	33.64
1140	12" dia.	"	10.25	29.00	39.25
2000	Pipe piles, non-filled				
2020	8" dia.	L.F.	7.42	17.00	24.42
2040	10" dia.	"	7.63	20.00	27.63
2060	12" dia.	"	7.86	24.50	32.36
2520	Splice				
2540	8" dia.	EA.	81.00	67.00	148
2560	10" dia.	"	81.00	71.00	152
2580	12" dia.	"	100	80.00	180
2680	Standard point				
2700	8" dia.	EA.	81.00	78.00	159
2740	10" dia.	"	81.00	88.00	169
2760	12" dia.	"	100	140	240
2880	Heavy duty point				
2900	8" dia.	EA.	100	73.00	173
2920	10" dia.	"	100	88.00	188
2940	12" dia.	"	140	120	260
02455.80	**WOOD AND TIMBER PILES**				
0080	Treated wood piles, 12" butt, 8" tip				
0100	25' long	L.F.	13.25	8.37	21.62
0110	30' long	"	11.25	8.94	20.19

		UNIT	LABOR	MAT.	TOTAL
02455.80	**WOOD AND TIMBER PILES, Cont'd...**				
0120	35' long	L.F.	9.54	8.94	18.48
0125	40' long	"	8.35	8.94	17.29
02465.50	**PRESTRESSED PILING**				
0980	Prestressed concrete piling, less than 60' long				
1000	10" sq.	L.F.	5.56	12.00	17.56
1002	12" sq.	"	5.81	16.50	22.31
1480	Straight cylinder, less than 60' long				
1500	12" dia.	L.F.	6.07	15.50	21.57
1540	14" dia.	"	6.21	20.75	26.96
02510.10	**WELLS**				
0980	Domestic water, drilled and cased				
1000	4" dia.	L.F.	67.00	27.75	94.75
1020	6" dia.	"	74.00	30.50	105
02510.40	**DUCTILE IRON PIPE**				
0990	Ductile iron pipe, cement lined, slip-on joints				
1000	4"	L.F.	6.19	13.50	19.69
1010	6"	"	6.56	16.25	22.81
1020	8"	"	6.97	21.25	28.22
1190	Mechanical joint pipe				
1200	4"	L.F.	8.58	16.00	24.58
1210	6"	"	9.29	19.00	28.29
1220	8"	"	10.25	25.00	35.25
1480	Fittings, mechanical joint				
1500	90 degree elbow				
1520	4"	EA.	27.00	190	217
1540	6"	"	31.25	240	271
1560	8"	"	40.50	350	391
1700	45 degree elbow				
1720	4"	EA.	27.00	160	187
1740	6"	"	31.25	220	251
1760	8"	"	40.50	300	341
02510.60	**PLASTIC PIPE**				
0110	PVC, class 150 pipe				
0120	4" dia.	L.F.	5.57	3.52	9.09
0130	6" dia.	"	6.02	6.66	12.68
0140	8" dia.	"	6.37	10.50	16.87
0165	Schedule 40 pipe				
0170	1-1/2" dia.	L.F.	2.38	0.86	3.24
0180	2" dia.	"	2.53	1.28	3.81
0185	2-1/2" dia.	"	2.70	1.95	4.65
0190	3" dia.	"	2.90	2.66	5.56
0200	4" dia.	"	3.38	3.76	7.14
0210	6" dia.	"	4.06	7.07	11.13
0240	90 degree elbows				
0250	1"	EA.	6.76	0.84	7.60
0260	1-1/2"	"	6.76	1.60	8.36
0270	2"	"	7.38	2.51	9.89
0280	2-1/2"	"	8.12	7.67	15.79
0290	3"	"	9.02	9.17	18.19
0300	4"	"	10.25	16.50	26.75
0310	6"	"	13.50	52.00	65.50

		UNIT	LABOR	MAT.	TOTAL
02510.60	**PLASTIC PIPE, Cont'd...**				
0500	Couplings				
0510	1"	EA.	6.76	0.68	7.44
0520	1-1/2"	"	6.76	0.97	7.73
0530	2"	"	7.38	1.50	8.88
0540	2-1/2"	"	8.12	3.32	11.44
0550	3"	"	9.02	5.20	14.22
0560	4"	"	10.25	7.52	17.77
0580	6"	"	13.50	23.75	37.25
02530.20	**VITRIFIED CLAY PIPE**				
0100	Vitrified clay pipe, extra strength				
1020	6" dia.	L.F.	10.25	5.21	15.46
1040	8" dia.	"	10.50	6.24	16.74
1050	10" dia.	"	11.25	9.57	20.82
02530.30	**MANHOLES**				
0100	Precast sections, 48" dia.				
0110	Base section	EA.	190	330	520
0120	1'0" riser	"	150	93.00	243
0130	1'4" riser	"	160	110	270
0140	2'8" riser	"	170	170	340
0150	4'0" riser	"	190	310	500
0160	2'8" cone top	"	220	200	420
0170	Precast manholes, 48" dia.				
0180	4' deep	EA.	450	610	1,060
0200	6' deep	"	560	930	1,490
0250	7' deep	"	640	1,060	1,700
0260	8' deep	"	740	1,190	1,930
0280	10' deep	"	890	1,340	2,230
1000	Cast-in-place, 48" dia., with frame and cover				
1100	5' deep	EA.	1,120	570	1,690
1120	6' deep	"	1,270	750	2,020
1140	8' deep	"	1,490	1,090	2,580
1160	10' deep	"	1,780	1,270	3,050
1480	Brick manholes, 48" dia. with cover, 8" thick				
1500	4' deep	EA.	510	600	1,110
1501	6' deep	"	560	760	1,320
1505	8' deep	"	630	980	1,610
1510	10' deep	"	720	1,210	1,930
4200	Frames and covers, 24" diameter				
4210	300 lb	EA.	40.50	350	391
4220	400 lb	"	45.00	370	415
4980	Steps for manholes				
5000	7" x 9"	EA.	8.12	12.00	20.12
5020	8" x 9"	"	9.02	15.25	24.27
02530.40	**SANITARY SEWERS**				
0980	Clay				
1000	6" pipe	L.F.	7.43	5.44	12.87
2980	PVC				
3000	4" pipe	L.F.	5.57	2.49	8.06
3010	6" pipe	"	5.87	4.99	10.86

		UNIT	LABOR	MAT.	TOTAL
02540.10	**DRAINAGE FIELDS**				
0080	Perforated PVC pipe, for drain field				
0100	4" pipe	L.F.	4.95	1.88	6.83
0120	6" pipe	"	5.31	3.53	8.84
02540.50	**SEPTIC TANKS**				
0980	Septic tank, precast concrete				
1000	1000 gals	EA.	370	750	1,120
1200	2000 gals	"	560	2,420	2,980
1310	Leaching pit, precast concrete, 72" diameter				
1320	3' deep	EA.	280	690	970
1340	6' deep	"	320	1,210	1,530
1360	8' deep	"	370	1,540	1,910
02630.70	**UNDERDRAIN**				
1480	Drain tile, clay				
1500	6" pipe	L.F.	4.95	4.52	9.47
1520	8" pipe	"	5.18	7.21	12.39
1580	Porous concrete, standard strength				
1600	6" pipe	L.F.	4.95	4.07	9.02
1620	8" pipe	"	5.18	4.40	9.58
1800	Corrugated metal pipe, perforated type				
1810	6" pipe	L.F.	5.57	5.64	11.21
1820	8" pipe	"	5.87	6.66	12.53
1980	Perforated clay pipe				
2000	6" pipe	L.F.	6.37	5.44	11.81
2020	8" pipe	"	6.56	7.29	13.85
2480	Drain tile, concrete				
2500	6" pipe	L.F.	4.95	3.23	8.18
2520	8" pipe	"	5.18	5.02	10.20
4980	Perforated rigid PVC underdrain pipe				
5000	4" pipe	L.F.	3.71	1.79	5.50
5100	6" pipe	"	4.46	3.28	7.74
5150	8" pipe	"	4.95	4.66	9.61
6980	Underslab drainage, crushed stone				
7000	3" thick	S.F.	0.74	0.24	0.98
7120	4" thick	"	0.85	0.30	1.15
7140	6" thick	"	0.92	0.37	1.29
7180	Plastic filter fabric for drain lines	"	0.40	0.15	0.55
02740.20	**ASPHALT SURFACES**				
0050	Asphalt wearing surface, for flexible pavement				
0100	1" thick	S.Y.	2.22	3.55	5.77
0120	1-1/2" thick	"	2.67	5.36	8.03
1000	Binder course				
1010	1-1/2" thick	S.Y.	2.47	5.08	7.55
1030	2" thick	"	3.03	6.75	9.78
2000	Bituminous sidewalk, no base				
2020	2" thick	S.Y.	2.62	7.74	10.36
2040	3" thick	"	2.78	11.50	14.28
02750.10	**CONCRETE PAVING**				
1080	Concrete paving, reinforced, 5000 psi concrete				
2000	6" thick	S.Y.	21.00	23.75	44.75
2005	7" thick	"	22.25	27.75	50.00
2010	8" thick	"	23.75	31.75	55.50

		UNIT	LABOR	MAT.	TOTAL
02810.40	**LAWN IRRIGATION**				
0480	Residential system, complete				
0500	Minimum	ACRE			15,940
0520	Maximum	"			30,340
02820.10	**CHAIN LINK FENCE**				
0230	Chain link fence, 9 ga., galvanized, with posts 10' o.c.				
0250	4' high	L.F.	2.90	6.71	9.61
0260	5' high	"	3.69	8.97	12.66
0270	6' high	"	5.07	10.25	15.32
1161	Fabric, galvanized chain link, 2" mesh, 9 ga.				
1163	4' high	L.F.	1.35	3.65	5.00
1164	5' high	"	1.62	4.46	6.08
1165	6' high	"	2.03	6.24	8.27
02820.20	**WOOD FENCE**				
0010	4X4 posts w/2x4 horizontals, 4' -8' high				
0020	Cedar, 1x4 planks, picket	S.F.	1.01	1.93	2.94
0030	1x6 planks, privacy	"	0.81	3.07	3.88
0040	1x8 planks, privacy	"	0.73	2.80	3.53
0050	Redwood, 1x4 planks, picket	"	1.01	1.57	2.58
0060	1x6 planks, privacy	"	0.81	2.91	3.72
0070	1x8 planks	"	0.73	2.54	3.27
0080	Treated pine, 1x4 planks, picket	"	1.01	1.54	2.55
0100	1x6 planks, privacy	"	0.81	1.51	2.32
0110	1x8 planks, privacy	"	0.73	1.48	2.21
0210	4' high				
0220	Composite, 1x4 planks, picket	S.F.	1.26	1.69	2.95
0230	1x6 planks, privacy	"	1.01	2.91	3.92
0240	1x8 planks	"	0.84	2.54	3.38
0250	Vinyl, 1x4 planks, picket	"	1.26	1.51	2.77
0260	1x6 planks, privacy	"	1.01	2.68	3.69
0270	1x8 planks, privacy	"	0.84	2.63	3.47
0280	Gate, cedar or redwood, 3' w, 1x4 planks, picket	EA.	33.75	26.50	60.25
0290	1x6 planks, privacy	"	40.50	73.00	114
0300	1x8 planks, privacy	"	45.00	83.00	128
02840.40	**PARKING BARRIERS**				
3010	Bollard, conc. filled, 8' long				
3020	6" dia.	EA.	34.75	410	445
3030	8" dia.	"	43.50	610	654
3040	12" dia.	"	52.00	780	832
02880.70	**RECREATIONAL COURTS**				
1000	Walls, galvanized steel				
1020	8' high	L.F.	8.12	14.50	22.62
1040	10' high	"	9.02	17.25	26.27
1060	12' high	"	10.75	19.75	30.50
1200	Vinyl coated				
1220	8' high	L.F.	8.12	13.75	21.87
1240	10' high	"	9.02	17.00	26.02
1260	12' high	"	10.75	18.75	29.50
2010	Gates, galvanized steel				
2200	Single, 3' transom				
2210	3'x7'	EA.	200	340	540
2220	4'x7'	"	230	360	590

		UNIT	LABOR	MAT.	TOTAL
02880.70	**RECREATIONAL COURTS, Cont'd...**				
2230	5'x7'	EA.	270	490	760
2240	6'x7'	"	320	530	850
2400	Vinyl coated				
2405	Single, 3' transom				
2410	3'x7'	EA.	200	660	860
2420	4'x7'	"	230	720	950
2430	5'x7'	"	270	720	990
2440	6'x7'	"	320	740	1,060
02910.10	**TOPSOIL**				
0005	Spread topsoil, with equipment				
0010	Minimum	C.Y.	12.00		12.00
0020	Maximum	"	15.00		15.00
0080	By hand				
0100	Minimum	C.Y.	40.50		40.50
0110	Maximum	"	51.00		51.00
0980	Area preparation for seeding (grade, rake and clean)				
1000	Square yard	S.Y.	0.32		0.32
1020	By acre	ACRE	1,620		1,620
2000	Remove topsoil and stockpile on site				
2020	4" deep	C.Y.	9.92		9.92
2040	6" deep	"	9.16		9.16
2200	Spreading topsoil from stock pile				
2220	By loader	C.Y.	10.75		10.75
2240	By hand	"	120		120
2260	Top dress by hand	S.Y.	1.19		1.19
2280	Place imported top soil				
2300	By loader				
2320	4" deep	S.Y.	1.19		1.19
2340	6" deep	"	1.32		1.32
2360	By hand				
2370	4" deep	S.Y.	4.51		4.51
2380	6" deep	"	5.07		5.07
5980	Plant bed preparation, 18" deep				
6000	With backhoe/loader	S.Y.	2.97		2.97
6010	By hand	"	6.76		6.76
02920.10	**FERTILIZING**				
0080	Fertilizing (23#/1000 sf)				
0100	By square yard	S.Y.	0.13	0.03	0.16
0120	By acre	ACRE	660	160	820
2980	Liming (70#/1000 sf)				
3000	By square yard	S.Y.	0.17	0.03	0.20
3020	By acre	ACRE	880	160	1,040
02920.30	**SEEDING**				
0980	Mechanical seeding, 175 lb/acre				
1000	By square yard	S.Y.	0.10	0.19	0.29
1020	By acre	ACRE	530	770	1,300
2040	450 lb/acre				
2060	By square yard	S.Y.	0.13	0.49	0.62
2080	By acre	ACRE	660	1,910	2,570
5980	Seeding by hand, 10 lb per 100 s.y.				
6000	By square yard	S.Y.	0.13	0.55	0.68

		UNIT	LABOR	MAT.	TOTAL
02920.30	**SEEDING, Cont'd...**				
6010	By acre	ACRE	680	2,130	2,810
8010	Reseed disturbed areas	S.F.	0.20	0.05	0.25
02935.10	**SHRUB & TREE MAINTENANCE**				
1000	Moving shrubs on site				
1220	3' high	EA.	40.50		40.50
1240	4' high	"	45.00		45.00
2000	Moving trees on site				
3060	6' high	EA.	49.50		49.50
3080	8' high	"	56.00		56.00
3100	10' high	"	74.00		74.00
3110	Palm trees				
3140	10' high	EA.	74.00		74.00
3144	40' high	"	450		450
02935.30	**WEED CONTROL**				
1000	Weed control, bromicil, 15 lb./acre, wettable powder	ACRE	200	290	490
1100	Vegetation control, by application of plant killer	S.Y.	0.16	0.02	0.18
1200	Weed killer, lawns and fields	"	0.08	0.24	0.32
02945.20	**LANDSCAPE ACCESSORIES**				
0100	Steel edging, 3/16" x 4"	L.F.	0.50	0.64	1.14
0200	Landscaping stepping stones, 15"x15", white	EA.	2.03	5.83	7.86
6000	Wood chip mulch	C.Y.	27.00	40.75	67.75
6010	2" thick	S.Y.	0.81	2.49	3.30
6020	4" thick	"	1.16	4.70	5.86
6030	6" thick	"	1.47	7.03	8.50
6200	Gravel mulch, 3/4" stone	C.Y.	40.50	32.25	72.75
6300	White marble chips, 1" deep	S.F.	0.40	0.64	1.04
6980	Peat moss				
7000	2" thick	S.Y.	0.90	3.46	4.36
7020	4" thick	"	1.35	6.66	8.01
7030	6" thick	"	1.69	10.25	11.94
7980	Landscaping timbers, treated lumber				
8000	4" x 4"	L.F.	1.35	1.35	2.70
8020	6" x 6"	"	1.45	2.69	4.14
8040	8" x 8"	"	1.69	4.40	6.09

		UNIT	COST
02999.10	**DEMOLITION**		
3200	Selective Building Removals		
3210	No Cutting or Disposal Included		
3220	Concrete - Hand Work		
3230	8" Walls - Reinforced	S.F.	14.00
3240	Non - Reinforced	"	9.29
3250	12" Walls - Reinforced	"	22.25
3260	12" Footings x 24" wide	L.F.	20.25
3270	x 36" wide	"	28.00
3280	16" Footings x 24" wide	S.F.	37.25
3290	6" Structural Slab - Reinforced	"	8.58
3300	8" Structural Slab - Reinforced	"	10.50
3310	4" Slab on Ground - Reinforced	"	3.98
3320	Non - Reinforced	"	2.97
3330	6" Slab on Ground - Reinforced	"	4.74
3340	Non - Reinforced	"	3.54
3350	Stairs - Reinforced	"	14.75
3360	Masonry - Hand Work		
3370	4" Brick or Stone Walls	S.F.	2.68
3380	4" Brick and 8" Backup Block or Tile	"	4.95
3390	4" Block or Tile Partitions	"	2.62
3400	6" Block or Tile Partitions	"	2.86
3410	8" Block or Tile Partitions	"	3.28
3420	12" Block or Tile Partitions	"	4.13
3430	Miscellaneous - Hand Work		
3440	Acoustical Ceilings - Attached	S.F.	0.76
3450	Suspended (Including Grid)	"	0.43
3460	Asbestos - Pipe	L.F.	56.00
3470	Ceilings and Walls	S.F.	18.50
3480	Columns and Beams	"	44.50
3490	Tile Flooring	"	2.08
3500	Cabinets and Tops	L.F.	12.50
3510	Carpet	S.F.	0.38
3520	Ceramic and Quarry Tile	"	1.48
3530	Doors and Frames - Metal	EA.	66.00
3540	Wood	"	56.00
3550	Drywall Ceilings - Attached	S.F.	0.93
3560	Drywall on Wood or Metal Studs - 2 Sides	"	1.04
3570	Paint Removal - Doors and Windows	"	0.93
3580	Walls	"	0.76
3590	Plaster Ceilings - Attached (Including Iron)	"	1.32
3600	on Wood or Metal Studs	"	1.26
3610	Roofing - Builtup	"	1.26
3620	Shingles - Asphalt and Wood	"	0.43
3630	Terrazzo Flooring	"	2.08
3640	Vinyl Flooring	"	0.43
3650	Wall Coverings	"	0.82
3660	Windows	EA.	44.50
3670	Wood Flooring	S.F.	0.43
4300	Site Removals (Including Loading)		
4310	4" Concrete Walks - Labor Only (Non-Reinforced)	S.F.	2.97
4320	Machine Only (Non-Reinforced)	"	0.98
4330	6" Concrete Drives - Labor Only (Non-Reinforced)	"	3.71

		UNIT	COST
02999.10	**DEMOLITION, Cont'd...**		
4340	Machine Only (Reinforced)	S.F.	1.32
4350	6" x 18" Concrete Curb - Machine	L.F.	2.53
4360	Curb and Gutter - Machine	"	3.29
4370	2" Asphalt - Machine	S.F.	0.57
4380	Fencing - 8' Hand	L.F.	3.40
02999.20	**EARTHWORK**		
1000	GRADING - Hand - 4" - Site	S.F.	0.34
1010	Hand – 4" Building	"	0.57
2000	EXCAVATION - Hand - Open - Soft (Sand)	C.Y.	44.50
2010	Medium (Clay)	"	56.00
2020	Hard (Shale)	"	89.00
2030	Add for Trench or Pocket	PCT.	15.00
3000	BACK FILL - Hand - Not Comp. (Site Borrow)	C.Y.	20.25
4000	BACK FILL - Hand - Comp.(Site Borrow)		
4010	12" Lifts - Building - No Machine	C.Y.	34.25
4020	With Machine	"	28.00
4030	18" Lifts - Building - No Machine	"	31.75
4040	With Machine	"	26.25
02999.50	**DRAINAGE**		
4000	BUILDING FOUNDATION DRAINAGE		
4010	4" Clay Pipe	L.F.	8.81
4020	4" Plastic Pipe - Perforated	"	7.10
4030	6" Clay Pipe	"	10.00
4040	6" Plastic Pipe - Perforated	"	11.29
4050	Add for Porous Surround - 2' x 2'	"	12.98
02999.60	**PAVEMENT, CURBS AND WALKS**		
2000	CURBS AND GUTTERS		
2100	Concrete - Cast in Place (Machine Placed)		
2110	Curb - 6" x 12"	L.F.	20.18
2120	6" x 18"	"	23.88
2130	6" x 24"	"	27.22
2140	6" x 30"	"	31.17
2150	Curb and Gutter - 6" x 12"	"	29.33
2160	6" x 18"	"	34.75
2170	6" x 24"	"	40.00
2180	Add for Hand Placed	"	9.29
2190	Add for 2 #5 Reinf. Rods	"	4.29
2191	Add for Curves and Radius Work	PCT.	40.00
2200	Concrete Precast - 6" x 10" x 8"	L.F.	22.11
2210	6" x 9" x 8"	"	19.84
2250	6" x 8"	"	17.57
2300	Bituminous - 6" x 8"	"	7.81
2400	Granite - 6" x 16"	"	64.00
2500	Timbers - Treated - 6" x 6"	"	14.73
2600	Plastic - 6" x 6"	"	19.11
3000	WALKS		
3010	Bituminous - 1 1/2" with 4" Sand Base	S.F.	3.87
3020	2" with 4" Sand Base	"	4.28
3030	Concrete - 4" - Broom Finish	"	7.08
3040	5" - Broom Finish	"	8.06
3050	6" - Broom Finish	"	9.13

		UNIT	COST
02999.60	**PAVEMENT, CURBS AND WALKS, Cont'd...**		
3060	Add for 6" x 6", 10 - 10 Mesh	S.F.	1.66
3070	Add for 4" Sand Base	"	1.42
3080	Add for Exposed Aggregate	"	2.24
3090	Crushed Rock - 4"	"	1.53
4000	Brick - 4" - with 2" Sand Cushion	"	17.31
4010	4" - with 2" Mortar Setting Bed	"	20.47
4020	Flagstone – 1 1/4" - with 4" Sand Cushion	"	28.30
4030	1¼" - with 2" Mortar Setting Bed	"	31.54
4040	Precast Block - 1" Colored with 4" Sand Cushion	"	7.93
4050	2" Colored with 4" Sand Cushion	"	9.19
4060	Wood - 2" Boards on 6" x 6" Timbers	"	14.01
4070	2" Boards on 4" x 4" Timbers	"	11.81
4080	Slate - 1 1/4" - with 2" Mortar Setting Bed	"	38.00

TABLE OF CONTENTS   PAGE

		UNIT	LABOR	MAT.	TOTAL
03100.03	**FORMWORK ACCESSORIES**				
1000	Column clamps				
1010	Small, adjustable, 24"x24"	EA.			56.00
1020	Medium 36"x36"	"			60.00
1030	Large 60"x60"	"			62.00
2000	Forming hangers				
2010	Iron 14 ga.	EA.			1.81
2020	22 ga.	"			1.81
3000	Snap ties				
3010	Short-end with washers, 6" long	EA.			1.15
3020	12" long	"			1.38
4000	18" long	"			1.59
4010	24" long	"			1.76
4020	Long-end with washers, 6' long	"			1.29
4030	12" long	"			1.45
4040	18" long	"			1.62
4050	24" long	"			1.92
5000	Stakes				
5010	Round, pre-drilled holes, 12" long	EA.			3.79
5020	18" long	"			4.89
5030	24" long	"			6.32
5040	30" long	"			8.25
5050	36" long	"			9.73
5060	48" long	"			13.00
5070	I beam type, 12" long	"			3.53
6000	18" long	"			4.02
6010	24" long	"			5.94
6020	30" long	"			7.20
6030	36" long	"			8.58
7000	48" long	"			10.75
7010	Taper ties				
7020	50K, 1-1/4" to 1", 35" long	EA.			120
8000	45" long	"			160
8010	55" long	"			170
8020	Walers				
8030	5" deep, 4' long	EA.			190
8040	8' long	"			240
8050	12' long	"			400
9000	16' long	"			520
9010	8" deep, 4' long	"			250
9020	8' long	"			560
9030	12' long	"			660
9040	16' long	"			1,050
03110.05	**BEAM FORMWORK**				
1000	Beam forms, job built				
1020	Beam bottoms				
1040	1 use	S.F.	8.70	4.14	12.84
1080	3 uses	"	8.03	1.87	9.90
1120	5 uses	"	7.45	1.41	8.86
2000	Beam sides				
2020	1 use	S.F.	5.80	2.95	8.75
2060	3 uses	"	5.22	1.54	6.76

		UNIT	LABOR	MAT.	TOTAL
03110.05	**BEAM FORMWORK, Cont'd...**				
2100	5 uses	S.F.	4.74	1.25	5.99
03110.15	**COLUMN FORMWORK**				
1000	Column, square forms, job built				
1020	8" x 8" columns				
1040	1 use	S.F.	10.50	3.25	13.75
1080	3 uses	"	9.66	1.48	11.14
1120	5 uses	"	9.00	1.15	10.15
1300	16" x 16" columns				
1320	1 use	S.F.	8.70	2.83	11.53
1360	3 uses	"	8.15	1.19	9.34
1390	5 uses	"	7.67	0.90	8.57
2000	Round fiber forms, 1 use				
2040	10" dia.	L.F.	10.50	4.29	14.79
2120	18" dia.	"	12.50	14.75	27.25
2180	36" dia.	"	15.75	33.75	49.50
03110.18	**CURB FORMWORK**				
0980	Curb forms				
0990	Straight, 6" high				
1000	1 use	L.F.	5.22	1.90	7.12
1040	3 uses	"	4.74	0.85	5.59
1080	5 uses	"	4.35	0.69	5.04
1090	Curved, 6" high				
2000	1 use	L.F.	6.52	2.05	8.57
2040	3 uses	"	5.80	0.99	6.79
2080	5 uses	"	5.32	0.83	6.15
03110.20	**ELEVATED SLAB FORMWORK**				
0100	Elevated slab formwork				
1000	Slab, with drop panels				
1020	1 use	S.F.	4.17	3.42	7.59
1060	3 uses	"	3.86	1.54	5.40
1100	5 uses	"	3.60	1.22	4.82
2000	Floor slab, hung from steel beams				
2020	1 use	S.F.	4.01	2.75	6.76
2060	3 uses	"	3.72	1.37	5.09
2100	5 uses	"	3.48	1.01	4.49
3000	Floor slab, with pans or domes				
3020	1 use	S.F.	4.74	4.90	9.64
3060	3 uses	"	4.35	2.97	7.32
3100	5 uses	"	4.01	2.45	6.46
9030	Equipment curbs, 12" high				
9035	1 use	L.F.	5.22	2.53	7.75
9060	3 uses	"	4.74	1.41	6.15
9100	5 uses	"	4.35	1.10	5.45
03110.25	**EQUIPMENT PAD FORMWORK**				
1000	Equipment pad, job built				
1020	1 use	S.F.	6.52	3.30	9.82
1040	2 uses	"	6.14	1.98	8.12
1060	3 uses	"	5.80	1.59	7.39

		UNIT	LABOR	MAT.	TOTAL
03110.35	**FOOTING FORMWORK**				
2000	Wall footings, job built, continuous				
2040	1 use	S.F.	5.22	1.56	6.78
2060	3 uses	"	4.74	0.90	5.64
2090	5 uses	"	4.35	0.69	5.04
3000	Column footings, spread				
3020	1 use	S.F.	6.52	1.65	8.17
3060	3 uses	"	5.80	0.88	6.68
3100	5 uses	"	5.22	0.67	5.89
03110.50	**GRADE BEAM FORMWORK**				
1000	Grade beams, job built				
1020	1 use	S.F.	5.22	2.45	7.67
1060	3 uses	"	4.74	1.07	5.81
1100	5 uses	"	4.35	0.74	5.09
03110.53	**PILE CAP FORMWORK**				
1500	Pile cap forms, job built				
1510	Square				
1520	1 use	S.F.	6.52	2.79	9.31
1560	3 uses	"	5.80	1.28	7.08
1600	5 uses	"	5.22	0.93	6.15
03110.55	**SLAB / MAT FORMWORK**				
3000	Mat foundations, job built				
3020	1 use	S.F.	6.52	2.44	8.96
3060	3 uses	"	5.80	1.03	6.83
3100	5 uses	"	5.22	0.70	5.92
3980	Edge forms				
3990	6" high				
4000	1 use	L.F.	4.74	2.45	7.19
4002	3 uses	"	4.35	1.03	5.38
4004	5 uses	"	4.01	0.70	4.71
4014	5 uses	"	4.35	0.64	4.99
5000	Formwork for openings				
5020	1 use	S.F.	10.50	3.30	13.80
5060	3 uses	"	8.70	1.59	10.29
5100	5 uses	"	7.45	1.02	8.47
03110.60	**STAIR FORMWORK**				
1000	Stairway forms, job built				
1020	1 use	S.F.	10.50	4.16	14.66
1040	3 uses	"	8.70	1.81	10.51
1060	5 uses	"	7.45	1.39	8.84
03110.65	**WALL FORMWORK**				
2980	Wall forms, exterior, job built				
3000	Up to 8' high wall				
3120	1 use	S.F.	5.22	2.70	7.92
3160	3 uses	"	4.74	1.32	6.06
3190	5 uses	"	4.35	0.99	5.34
3200	Over 8' high wall				
3220	1 use	S.F.	6.52	2.97	9.49
3240	3 uses	"	5.80	1.55	7.35
3290	5 uses	"	5.22	1.22	6.44
5000	Column pier and pilaster				
5020	1 use	S.F.	10.50	2.97	13.47

		UNIT	LABOR	MAT.	TOTAL
03110.65	**WALL FORMWORK, Cont'd...**				
5060	3 uses	S.F.	8.70	1.63	10.33
5090	5 uses	"	7.45	1.34	8.79
6980	Interior wall forms				
7000	Up to 8' high				
7020	1 use	S.F.	4.74	2.70	7.44
7060	3 uses	"	4.35	1.34	5.69
7100	5 uses	"	4.01	0.96	4.97
7200	Over 8' high				
7220	1 use	S.F.	5.80	2.97	8.77
7260	3 uses	"	5.22	1.55	6.77
7290	5 uses	"	4.74	1.23	5.97
9000	PVC form liner, per side, smooth finish				
9010	1 use	S.F.	4.35	6.95	11.30
9030	3 uses	"	4.01	3.23	7.24
9050	5 uses	"	3.48	1.99	5.47
03110.70	**INSULATED CONCRETE FORMS**				
0010	4" thick, straight	S.F.	1.63	4.35	5.98
0020	90° corner	"	1.63	4.34	5.97
0030	45° angle	"	1.63	4.75	6.38
1000	6" thick, straight	"	1.63	4.40	6.03
1010	90° corner	"	1.63	4.40	6.03
1020	45° angle	"	1.63	4.84	6.47
1030	6" thick, corbel ledge	"	1.63	5.50	7.13
1040	T-block	"	1.63	4.95	6.58
2000	8" thick, straight	"	1.63	4.62	6.25
2010	90° corner	"	1.63	4.40	6.03
2020	45° angle	"	1.63	5.06	6.69
2030	corbel ledge	"	1.63	5.66	7.29
2040	T-block	"	1.63	5.22	6.85
03110.90	**MISCELLANEOUS FORMWORK**				
1200	Keyway forms (5 uses)				
1220	2 x 4	L.F.	2.61	0.22	2.83
1240	2 x 6	"	2.90	0.33	3.23
1500	Bulkheads				
1510	Walls, with keyways				
1515	2 piece	L.F.	4.74	3.69	8.43
1520	3 piece	"	5.22	4.67	9.89
1560	Elevated slab, with keyway				
1570	2 piece	L.F.	4.35	4.23	8.58
1580	3 piece	"	4.74	6.23	10.97
1600	Ground slab, with keyway				
1620	2 piece	L.F.	3.72	4.37	8.09
1640	3 piece	"	4.01	5.34	9.35
2000	Chamfer strips				
2020	Wood				
2040	1/2" wide	L.F.	1.16	0.20	1.36
2060	3/4" wide	"	1.16	0.27	1.43
2070	1" wide	"	1.16	0.36	1.52
2100	PVC				
2120	1/2" wide	L.F.	1.16	0.93	2.09
2140	3/4" wide	"	1.16	1.02	2.18

		UNIT	LABOR	MAT.	TOTAL
03110.90	**MISCELLANEOUS FORMWORK, Cont'd...**				
2160	1" wide	L.F.	1.16	1.47	2.63
2170	Radius				
2180	1"	L.F.	1.24	1.10	2.34
2200	1-1/2"	"	1.24	1.98	3.22
3000	Reglets				
3020	Galvanized steel, 24 ga.	L.F.	2.08	1.43	3.51
5000	Metal formwork				
5020	Straight edge forms				
5080	8" high	L.F.	3.72	26.25	29.97
5300	Curb form, S-shape				
5310	12" x				
5340	2'	L.F.	6.96	42.00	48.96
5380	3'	"	5.80	49.75	55.55
03210.05	**BEAM REINFORCING**				
0980	Beam-girders				
1000	#3 - #4	TON	1,450	1,270	2,720
1011	#7 - #8	"	970	1,060	2,030
1018	Galvanized				
1020	#3 - #4	TON	1,450	2,160	3,610
1031	#7 - #8	"	970	1,970	2,940
2000	Bond Beams				
2100	#3 - #4	TON	1,940	1,270	3,210
2120	#7 - #8	"	1,290	1,060	2,350
2200	Galvanized				
2210	#3 - #4	TON	1,940	2,070	4,010
2230	#7 - #8	"	1,290	1,970	3,260
03210.15	**COLUMN REINFORCING**				
0980	Columns				
1000	#3 - #4	TON	1,660	1,270	2,930
1015	#7 - #8	"	1,160	1,060	2,220
1100	Galvanized				
1200	#3 - #4	TON	1,660	2,160	3,820
1320	#7 - #8	"	1,160	1,970	3,130
03210.20	**ELEVATED SLAB REINFORCING**				
0980	Elevated slab				
1000	#3 - #4	TON	730	1,270	2,000
1040	#7 - #8	"	580	1,060	1,640
1980	Galvanized				
2000	#3 - #4	TON	730	2,070	2,800
2040	#7 - #8	"	580	1,970	2,550
03210.25	**EQUIP. PAD REINFORCING**				
0980	Equipment pad				
1000	#3 - #4	TON	1,160	1,270	2,430
1040	#7 - #8	"	970	1,060	2,030
03210.35	**FOOTING REINFORCING**				
1000	Footings				
1010	Grade 50				
1020	#3 - #4	TON	970	1,270	2,240
1040	#7 - #8	"	730	1,060	1,790
1055	Grade 60				
1060	#3 - #4	TON	970	1,270	2,240

		UNIT	LABOR	MAT.	TOTAL
03210.35	**FOOTING REINFORCING, Cont'd...**				
1074	#7 - #8	TON	730	1,060	1,790
4980	Straight dowels, 24" long				
5000	1" dia. (#8)	EA.	5.80	3.94	9.74
5040	3/4" dia. (#6)	"	5.80	3.55	9.35
5050	5/8" dia. (#5)	"	4.84	3.06	7.90
5060	1/2" dia. (#4)	"	4.14	2.31	6.45
03210.45	**FOUNDATION REINFORCING**				
0980	Foundations				
1000	#3 - #4	TON	970	1,270	2,240
1040	#7 - #8	"	730	1,060	1,790
1380	Galvanized				
1400	#3 - #4	TON	970	2,160	3,130
1420	#7 - #8	"	730	1,970	2,700
03210.50	**GRADE BEAM REINFORCING**				
0980	Grade beams				
1000	#3 - #4	TON	890	1,270	2,160
1040	#7 - #8	"	680	1,060	1,740
1090	Galvanized				
1100	#3 - #4	TON	890	2,160	3,050
1140	#7 - #8	"	680	1,970	2,650
03210.53	**PILE CAP REINFORCING**				
0980	Pile caps				
1000	#3 - #4	TON	1,450	1,270	2,720
1040	#7 - #8	"	1,160	1,060	2,220
1090	Galvanized				
1100	#3 - #4	TON	1,450	2,160	3,610
1140	#7 - #8	"	1,160	1,970	3,130
03210.55	**SLAB / MAT REINFORCING**				
0980	Bars, slabs				
1000	#3 - #4	TON	970	1,270	2,240
1040	#7 - #8	"	730	1,060	1,790
1980	Galvanized				
2000	#3 - #4	TON	970	2,160	3,130
2020	#5 - #6	"	830	2,040	2,870
2040	#7 - #8	"	730	1,970	2,700
5000	Wire mesh, slabs				
5010	Galvanized				
5015	4x4				
5020	W1.4xW1.4	S.F.	0.38	0.31	0.69
5040	W2.0xW2.0	"	0.41	0.40	0.81
5060	W2.9xW2.9	"	0.44	0.57	1.01
5080	W4.0xW4.0	"	0.48	0.84	1.32
5090	6x6				
5100	W1.4xW1.4	S.F.	0.29	0.29	0.58
5120	W2.0xW2.0	"	0.32	0.40	0.72
5140	W2.9xW2.9	"	0.34	0.55	0.89
5160	W4.0xW4.0	"	0.38	0.59	0.97
5170	Standard				
5175	2x2				
5180	W.9xW.9	S.F.	0.38	0.31	0.69
5185	4x4				

		UNIT	LABOR	MAT.	TOTAL
03210.55	**SLAB / MAT REINFORCING, Cont'd...**				
5190	W1.4xW1.4	S.F.	0.38	0.20	0.58
5500	W4.0xW4.0	"	0.48	0.57	1.05
5580	6x6				
5600	W1.4xW1.4	S.F.	0.29	0.13	0.42
6020	W4.0xW4.0	"	0.38	0.38	0.76
03210.60	**STAIR REINFORCING**				
0980	Stairs				
1000	#3 - #4	TON	1,160	1,270	2,430
1020	#5 - #6	"	970	1,120	2,090
1980	Galvanized				
2000	#3 - #4	TON	1,160	2,160	3,320
2020	#5 - #6	"	970	2,040	3,010
03210.65	**WALL REINFORCING**				
0980	Walls				
1000	#3 - #4	TON	830	1,270	2,100
1040	#7 - #8	"	650	1,060	1,710
1980	Galvanized				
2000	#3 - #4	TON	830	2,160	2,990
2040	#7 - #8	"	650	1,970	2,620
8980	Masonry wall (horizontal)				
9000	#3 - #4	TON	2,320	1,270	3,590
9020	#5 - #6	"	1,940	1,120	3,060
9030	Galvanized				
9040	#3 - #4	TON	2,320	2,160	4,480
9060	#5 - #6	"	1,940	2,040	3,980
9180	Masonry wall (vertical)				
9200	#3 - #4	TON	2,900	1,270	4,170
9220	#5 - #6	"	2,320	1,120	3,440
9230	Galvanized				
9240	#3 - #4	TON	2,900	2,160	5,060
9260	#5 - #6	"	2,320	2,040	4,360
03250.40	**CONCRETE ACCESSORIES**				
1000	Expansion joint, poured				
1010	Asphalt				
1020	1/2" x 1"	L.F.	0.81	0.79	1.60
1040	1" x 2"	"	0.88	2.47	3.35
1060	Liquid neoprene, cold applied				
1080	1/2" x 1"	L.F.	0.82	2.91	3.73
1100	1" x 2"	"	0.90	12.00	12.90
03300.10	**CONCRETE ADMIXTURES**				
1000	Concrete admixtures				
1020	Water reducing admixture	GAL			9.35
1040	Set retarder	"			20.50
1060	Air entraining agent	"			8.25
03350.10	**CONCRETE FINISHES**				
0980	Floor finishes				
1000	Broom	S.F.	0.58		0.58
1020	Screed	"	0.50		0.50
1040	Darby	"	0.50		0.50
1060	Steel float	"	0.67		0.67
4000	Wall finishes				

		UNIT	LABOR	MAT.	TOTAL
03350.10	**CONCRETE FINISHES, Cont'd...**				
4020	Burlap rub, with cement paste	S.F.	0.67	0.11	0.78
4160	Break ties and patch holes	"	0.81		0.81
4170	Carborundum				
4180	Dry rub	S.F.	1.35		1.35
4200	Wet rub	"	2.03		2.03
5000	Floor hardeners				
5010	Metallic				
5020	Light service	S.F.	0.50	0.36	0.86
5040	Heavy service	"	0.67	1.10	1.77
5050	Non-metallic				
5060	Light service	S.F.	0.50	0.18	0.68
5080	Heavy service	"	0.67	0.77	1.44
03360.10	**PNEUMATIC CONCRETE**				
0100	Pneumatic applied concrete (gunite)				
1035	2" thick	S.F.	2.78	5.24	8.02
1040	3" thick	"	3.71	6.44	10.15
1060	4" thick	"	4.46	7.85	12.31
1980	Finish surface				
2000	Minimum	S.F.	2.61		2.61
2020	Maximum	"	5.22		5.22
03370.10	**CURING CONCRETE**				
1000	Sprayed membrane				
1010	Slabs	S.F.	0.08	0.05	0.13
1020	Walls	"	0.10	0.07	0.17
1025	Curing paper				
1030	Slabs	S.F.	0.10	0.07	0.17
2000	Walls	"	0.11	0.07	0.18
2010	Burlap				
2020	7.5 oz.	S.F.	0.13	0.06	0.19
2500	12 oz.	"	0.14	0.08	0.22
03380.05	**BEAM CONCRETE**				
0960	Beams and girders				
0980	2500# or 3000# concrete				
1010	By pump	C.Y.	73.00	110	183
1020	By hand buggy	"	40.50	110	151
4000	5000# concrete				
4020	By pump	C.Y.	73.00	120	193
4040	By hand buggy	"	40.50	120	161
9460	Bond beam, 3000# concrete				
9470	By pump				
9480	8" high				
9500	4" wide	L.F.	1.61	0.29	1.90
9520	6" wide	"	1.83	0.71	2.54
9530	8" wide	"	2.01	0.91	2.92
9540	10" wide	"	2.24	1.21	3.45
9550	12" wide	"	2.52	1.62	4.14
03380.15	**COLUMN CONCRETE**				
0980	Columns				
0990	2500# or 3000# concrete				
1010	By pump	C.Y.	67.00	110	177
3980	5000# concrete				

		UNIT	LABOR	MAT.	TOTAL
03380.15	**COLUMN CONCRETE, Cont'd...**				
4020	By pump	C.Y.	67.00	120	187
03380.25	**EQUIPMENT PAD CONCRETE**				
0960	Equipment pad				
0980	2500# or 3000# concrete				
1000	By chute	C.Y.	13.50	110	124
1020	By pump	"	58.00	110	168
1050	3500# or 4000# concrete				
1060	By chute	C.Y.	13.50	110	124
1080	By pump	"	58.00	110	168
1110	5000# concrete				
1120	By chute	C.Y.	13.50	120	134
1140	By pump	"	58.00	120	178
03380.35	**FOOTING CONCRETE**				
0980	Continuous footing				
0990	2500# or 3000# concrete				
1000	By chute	C.Y.	13.50	110	124
1010	By pump	"	50.00	110	160
4000	5000# concrete				
4010	By chute	C.Y.	13.50	120	134
4020	By pump	"	50.00	120	170
4980	Spread footing				
5000	2500# or 3000# concrete				
5010	Under 5 cy				
5020	By chute	C.Y.	13.50	100	114
5040	By pump	"	54.00	100	154
7200	5000# concrete				
7205	Under 5 c.y.				
7210	By chute	C.Y.	13.50	120	134
7220	By pump	"	54.00	120	174
03380.50	**GRADE BEAM CONCRETE**				
0960	Grade beam				
0980	2500# or 3000# concrete				
1000	By chute	C.Y.	13.50	100	114
1040	By pump	"	50.00	100	150
1060	By hand buggy	"	40.50	100	141
1150	5000# concrete				
1160	By chute	C.Y.	13.50	120	134
1190	By pump	"	50.00	120	170
1200	By hand buggy	"	40.50	120	161
03380.53	**PILE CAP CONCRETE**				
0970	Pile cap				
0980	2500# or 3000 concrete				
1000	By chute	C.Y.	13.50	110	124
1010	By pump	"	58.00	110	168
1020	By hand buggy	"	40.50	110	151
3980	5000# concrete				
4010	By chute	C.Y.	13.50	120	134
4020	By pump	"	58.00	120	178
4030	By hand buggy	"	40.50	120	161

		UNIT	LABOR	MAT.	TOTAL
03380.55	**SLAB / MAT CONCRETE**				
0960	Slab on grade				
0980	2500# or 3000# concrete				
1000	By chute	C.Y.	10.25	110	120
1020	By pump	"	28.75	110	139
1030	By hand buggy	"	27.00	110	137
3980	5000# concrete				
4010	By chute	C.Y.	10.25	120	130
4030	By pump	"	28.75	120	149
4040	By hand buggy	"	27.00	120	147
03380.58	**SIDEWALKS**				
6000	Walks, cast in place with wire mesh, base not incl.				
6010	4" thick	S.F.	1.35	1.52	2.87
6020	5" thick	"	1.62	2.06	3.68
6030	6" thick	"	2.03	2.54	4.57
03380.60	**STAIR CONCRETE**				
0960	Stairs				
0980	2500# or 3000# concrete				
1000	By chute	C.Y.	13.50	110	124
1030	By pump	"	58.00	110	168
1040	By hand buggy	"	40.50	110	151
2100	3500# or 4000# concrete				
2120	By chute	C.Y.	13.50	110	124
2160	By pump	"	58.00	110	168
2180	By hand buggy	"	40.50	110	151
4000	5000# concrete				
4010	By chute	C.Y.	13.50	120	134
4030	By pump	"	58.00	120	178
4040	By hand buggy	"	40.50	120	161
03380.65	**WALL CONCRETE**				
0940	Walls				
0960	2500# or 3000# concrete				
0980	To 4'				
1000	By chute	C.Y.	11.50	110	122
1010	By pump	"	62.00	110	172
1020	To 8'				
1040	By pump	C.Y.	67.00	110	177
2960	3500# or 4000# concrete				
2980	To 4'				
3000	By chute	C.Y.	11.50	110	122
3030	By pump	"	62.00	110	172
3060	To 8'				
3100	By pump	C.Y.	67.00	110	177
8480	Filled block (CMU)				
8490	3000# concrete, by pump				
8500	4" wide	S.F.	2.88	0.40	3.28
8510	6" wide	"	3.36	0.92	4.28
8520	8" wide	"	4.03	1.44	5.47
8530	10" wide	"	4.74	1.92	6.66
8540	12" wide	"	5.76	2.47	8.23
8560	Pilasters, 3000# concrete	C.F.	81.00	5.61	86.61
8700	Wall cavity, 2" thick, 3000# concrete	S.F.	2.69	1.04	3.73

		UNIT	LABOR	MAT.	TOTAL
03550.10	**CONCRETE TOPPINGS**				
1000	Gypsum fill				
1020	2" thick	S.F.	0.41	1.63	2.04
1040	2-1/2" thick	"	0.42	1.87	2.29
1060	3" thick	"	0.43	2.29	2.72
1080	3-1/2" thick	"	0.44	2.62	3.06
1100	4" thick	"	0.50	3.06	3.56
2000	Formboard				
2020	Mineral fiber board				
2040	1" thick	S.F.	1.01	1.46	2.47
2060	1-1/2" thick	"	1.16	3.85	5.01
2070	Cement fiber board				
2080	1" thick	S.F.	1.35	1.14	2.49
2100	1-1/2" thick	"	1.56	1.46	3.02
2110	Glass fiber board				
2120	1" thick	S.F.	1.01	1.80	2.81
2140	1-1/2" thick	"	1.16	2.44	3.60
4000	Poured deck				
4010	Vermiculite or perlite				
4020	1 to 4 mix	C.Y.	67.00	160	227
4040	1 to 6 mix	"	62.00	140	202
4050	Vermiculite or perlite				
4060	2" thick				
4080	1 to 4 mix	S.F.	0.42	1.46	1.88
4100	1 to 6 mix	"	0.38	1.06	1.44
4200	3" thick				
4220	1 to 4 mix	S.F.	0.62	2.00	2.62
4240	1 to 6 mix	"	0.57	1.58	2.15
6000	Concrete plank, lightweight				
6020	2" thick	S.F.	3.34	7.95	11.29
6040	2-1/2" thick	"	3.34	8.16	11.50
6080	3-1/2" thick	"	3.71	8.49	12.20
6100	4" thick	"	3.71	8.80	12.51
6500	Channel slab, lightweight, straight				
6520	2-3/4" thick	S.F.	3.34	6.41	9.75
6540	3-1/2" thick	"	3.34	6.60	9.94
6560	3-3/4" thick	"	3.34	7.12	10.46
6580	4-3/4" thick	"	3.71	9.02	12.73
7000	Gypsum plank				
7020	2" thick	S.F.	3.34	2.97	6.31
7040	3" thick	"	3.34	3.11	6.45
8000	Cement fiber, T and G planks				
8020	1" thick	S.F.	3.03	1.63	4.66
8040	1-1/2" thick	"	3.03	1.73	4.76
8060	2" thick	"	3.34	2.07	5.41
8080	2-1/2" thick	"	3.34	2.20	5.54
8100	3" thick	"	3.34	2.86	6.20
8120	3-1/2" thick	"	3.71	3.30	7.01
8140	4" thick	"	3.71	3.63	7.34
03730.10	**CONCRETE REPAIR**				
0090	Epoxy grout floor patch, 1/4" thick	S.F.	4.06	6.06	10.12
0100	Grout, epoxy, 2 component system	C.F.			290

		UNIT	LABOR	MAT.	TOTAL
03730.10	**CONCRETE REPAIR, Cont'd...**				
0110	Epoxy sand	BAG			19.75
0120	Epoxy modifier	GAL			130
0140	Epoxy gel grout	S.F.	40.50	2.95	43.45
0150	Injection valve, 1 way, threaded plastic	EA.	8.12	8.11	16.23
0155	Grout crack seal, 2 component	C.F.	40.50	680	721
0160	Grout, non shrink	"	40.50	70.00	111
0165	Concrete, epoxy modified				
0170	Sand mix	C.F.	16.25	110	126
0180	Gravel mix	"	15.00	88.00	103
0190	Concrete repair				
0195	Soffit repair				
0200	16" wide	L.F.	8.12	3.46	11.58
0210	18" wide	"	8.45	3.68	12.13
0220	24" wide	"	9.02	4.40	13.42
0230	30" wide	"	9.66	4.95	14.61
0240	32" wide	"	10.25	5.28	15.53
0245	Edge repair				
0250	2" spall	L.F.	10.25	1.65	11.90
0260	3" spall	"	10.75	1.65	12.40
0270	4" spall	"	11.00	1.76	12.76
0280	6" spall	"	11.25	1.81	13.06
0290	8" spall	"	12.00	1.92	13.92
0300	9" spall	"	13.50	1.98	15.48
0330	Crack repair, 1/8" crack	"	4.06	3.24	7.30
5000	Reinforcing steel repair				
5005	1 bar, 4 ft				
5010	#4 bar	L.F.	7.26	0.50	7.76
5012	#5 bar	"	7.26	0.69	7.95
5014	#6 bar	"	7.74	0.83	8.57
5016	#8 bar	"	7.74	1.51	9.25
5020	#9 bar	"	8.29	1.93	10.22
5030	#11 bar	"	8.29	3.02	11.31
7010	Form fabric, nylon				
7020	18" diameter	L.F.			13.00
7030	20" diameter	"			13.25
7040	24" diameter	"			22.00
7050	30" diameter	"			22.50
7060	36" diameter	"			25.75
7100	Pile repairs				
7105	Polyethylene wrap				
7108	30 mil thick				
7110	60" wide	S.F.	13.50	14.25	27.75
7120	72" wide	"	16.25	15.50	31.75
7125	60 mil thick				
7130	60" wide	S.F.	13.50	16.75	30.25
7140	80" wide	"	18.50	19.50	38.00
8010	Pile spall, average repair 3'				
8020	18" x 18"	EA.	33.75	44.00	77.75
8030	20" x 20"	"	40.50	59.00	99.50

03999.10	CONCRETE	UNIT	COST
1000	FOOTINGS (Incl. exc. with steel)		
1010	By L.F - Continuous 24" x 12" (3 # 4 Rods)	L.F.	27.00
1020	36" x 12" (4 # 4 Rods)	"	34.25
1030	20" x 10" (2 # 5 Rods)	"	22.75
1040	16" x 8" (2 # 4 Rods)	"	19.42
1050	Pad 24" x 24" x 12" (4 # 5 E.W.)	EA.	93.75
1060	36" x 36" x 14" (6 # 5 E.W.)	"	189
1070	48" x 48" x 16" (8 # 5 E.W.)	"	340
1100	By C.Y. - Continuous 24" x 12" (3 # 4 Rods)	C.Y.	340
1110	36" x 12" (4 # 4 Rods)	"	350
1120	20" x 10" (2 # 5 Rods)	"	370
1130	16" x 8" (2 # 4 Rods)	"	450
1140	Pad 24" x 24" x 12" (4 # 5 E.W.)	"	560
1150	36" x 36" x 14" (6 # 5 E.W.)	"	520
1160	48" x 48" x 16" (8 # 5 E.W.)	"	500
1200	WALLS (#5 Rods 12" O.C. - 1 Face)		
1210	By S.F. - 8" Wall (# 5 12" O.C. E.W.)	S.F.	19.42
1220	12" Wall (# 5 12" O.C. E.W.)	"	20.95
1230	16" Wall (# 5 12" O.C. E.W.)	"	20.95
1300	Add for Steel - 2 Faces - 12" Wall	"	2.49
1310	Add for Pilastered Wall - 24" O.C.	"	1.17
1320	Add for Retaining or Battered Type	"	3.51
1330	Add for Curved Walls	"	6.50
1340	By C.Y. - 8" Wall (# 5 12" O.C. E.W.)	C.Y.	620
1350	12" Wall (# 5 12" O.C. E.W.)	"	510
1360	16" Wall (# 5 12" O.C. E.W.)	"	470
1400	Add for Steel - 2 Faces - 12" Wall	"	68.75
1410	Add for Pilastered Wall - 24" O.C.	"	30.50
1420	Add for Retaining or Battered Type	"	94.25
1430	Add for Curved Walls	"	172
1500	COLUMNS		
1510	By L.F. - Square Cornered 8" x 8" (4 # 8 Rods)	L.F.	44.50
1520	12" x 12" (6 # 8 Rods)	"	75.50
1530	16" x 16" (6 # 10 Rods)	"	97.50
1540	20" x 20" (8 # 20 Rods)	"	141
1550	24" x 24" (10 # 11 Rods)	"	168
1600	Round 8" (4 # 8 Rods)	"	32.50
1610	12" (6 # 8 Rods)	"	51.00
1620	16" (6 # 10 Rods)	"	69.50
1630	20" (8 # 20 Rods)	"	97.50
1640	24" (10 # 11 Rods)	"	141
1700	By C.Y. - Sq Cornered 8" x 8" (4 # 8 Rods)	C.Y.	2,580
1710	12" x 12" (6 # 8 Rods)	"	2,380
1720	16" x 16" (6 # 10 Rods)	"	1,690
1730	20" x 20" (8 # 20 Rods)	"	1,500
1740	24" x 24" (10 # 11 Rods)	"	1,080
1800	Round 8" (4 # 8 Rods)	"	2,380
1810	12" (6 # 8 Rods)	"	1,580
1820	16" (6 # 10 Rods)	"	1,460
1830	20" (8 # 20 Rods)	"	1,330
1840	24" (10 # 11 Rods)	"	1,270
1900	BEAMS		

		UNIT	COST
03999.10	**CONCRETE, Cont'd...**		
1910	By L.F. - Spandrel 12" x 48" (33 # rebar)	L.F.	169
1920	12" x 42" (26 # Rein. Steel)	"	156
1930	12" x 36" (21 # Rein. Steel)	"	123
1940	12" x 30" (15 # Rein. Steel)	"	116
1950	8" x 48" (26 # Rein. Steel)	"	152
1960	8" x 42" (21 # Rein. Steel)	"	138
1970	8" x 36" (16 # Rein. Steel)	"	125
2000	Interior 16" x 30" (24 # rebar)	"	99.50
2010	16" x 24" (20 # Rein. Steel)	"	98.50
2020	12" x 30" (17 # Rein. Steel)	"	115
2030	12" x 24" (14 # Rein. Steel)	"	82.00
2040	12" x 16" (12 # Rein. Steel)	"	69.50
2050	8" x 24" (13 # Rein. Steel)	"	82.00
2060	8" x 16" (10 # Rein. Steel)	"	61.75
2100	By C.Y. - Spandrel 12" x 48" (33 # rebar)	C.Y.	1,020
2110	12" x 42" (26 # Rein. Steel)	"	1,080
2120	12" x 36" (21 # Rein. Steel)	"	1,140
2130	12" x 30" (15 # Rein. Steel)	"	1,210
2140	8" x 48" (26 # Rein. Steel)	"	1,210
2150	8" x 42" (21 # Rein. Steel)	"	1,320
2160	8" x 36" (16 # Rein. Steel)	"	1,380
2200	Interior 16" x 30" (24 # rebar)	"	660
2210	16" x 24" (20 # Rein. Steel)	"	990
2220	12" x 30" (17 # Rein. Steel)	"	1,080
2230	12" x 24" (14 # Rein. Steel)	"	1,060
2240	12" x 16" (12 # Rein. Steel)	"	1,200
2250	8" x 24" (13 # Rein. Steel)	"	1,330
2260	8" x 16" (10 # Rein. Steel)	"	1,400
2300	SLABS (With Reinf. Steel)		
2310	By S.F. - Solid		
2320	4" Thick	S.F.	11.98
2330	5" Thick	"	14.04
2340	6" Thick	"	15.44
2350	7" Thick	"	17.20
2360	8" Thick	"	16.49
2400	Deduct for Post Tensioned Slabs	"	1.03
2410	Pan	"	
2420	Joist -20" Pan - 10" x 2"	"	15.83
2430	12" x 2"	"	17.81
2440	30" Pan - 10" x 2 1/2"	"	15.96
2450	12" x 2 1/2"	"	17.81
2500	Dome -19" x 19" - 10" x 2"	"	17.49
2510	12" x 2"	"	18.05
2520	30" x 30" - 10" x 2 1/2"	"	15.40
2530	12" x 2 1/2"	"	17.10
2600	By C.Y. - Solid		
2610	4" Thick	C.Y.	920
2620	5" Thick	"	880
2630	6" Thick	"	850
2640	8" Thick	"	820
2700	COMBINED COLUMNS, BEAMS AND SLABS		
2710	20' Span	S.F.	24.25

		UNIT	COST
03999.10	**CONCRETE, Cont'd...**		
2720	30' Span	S.F.	25.75
2730	40' Span	"	27.00
2740	50' Span	"	29.50
2750	60' Span	"	49.50
2800	STAIRS (Including Landing)		
2810	By EA. - 4' Wide 10' Floor Heights 16 Risers	EA.	220
2820	5' Wide 10' Floor Heights 16 Risers	"	240
2830	6' Wide 10' Floor Heights 16 Risers	"	300
2840	By C.Y. - 4' Wide 10' Floor Heights 16 Risers	C.Y.	1,330
2850	5' Wide 10' Floor Heights 16 Risers	"	1,380
2860	6' Wide 10' Floor Heights 16 Risers	"	1,350
2900	SLABS ON GROUND		
2910	4" Concrete Slab		
2920	(6 6/10 - 10 Mesh - 5 1/2 Sack Concrete,		
2930	Trowel Finished, Cured & Truck Chuted)	S.F.	4.07
2940	Add per Inch of Concrete	"	0.69
2950	Add per Sack of Cement	"	0.17
2960	Deduct for Float Finish	"	0.09
2970	Deduct for Brush or Broom Finish	"	0.07
3000	Add for Runway and Buggied Concrete	"	0.26
3010	Add for Vapor Barrier (4 mil)	"	0.18
3020	Add for Sub - Floor Fill (4" sand/gravel)	"	0.58
3030	Add for Change to 6 6/8 - 8 Mesh	"	0.09
3040	Add for Change to 6 6/6 - 6 Mesh	"	0.18
3050	Add for Sloped Slab	"	0.25
3060	Add for Edge Strip (sidewalk area)	"	0.27
3100	Add for ½" Expansion Joint (20' O.C.)	"	0.09
3110	Add for Control Joints (keyed/ dep.)	"	0.20
3120	Add for Control Joints (joint filled)	"	0.21
3130	Add for Control Joints - Saw Cut (20' O.C.)	"	0.37
3140	Add for Floor Hardener (1 coat)	"	0.18
3150	Add for Exposed Aggregate - Washed Added	"	0.51
3200	Retarding Added	"	0.49
3210	Seeding Added	"	0.51
3220	Add for Light Weight Aggregates	"	0.51
3230	Add for Heavy Weight Aggregates	"	0.35
3240	Add for Winter Production Loss/Cost	"	0.44
3300	TOPPING SLABS		
3310	2" Concrete	S.F.	3.48
3320	3" Concrete	"	4.34
3330	No Mesh or Hoisting		
3400	PADS & PLATFORMS (Including Form Work & Reinforcing)		
3410	4"	S.F.	8.50
3420	6"	"	10.73
3500	PRECAST CONCRETE ITEMS		
3510	Curbs 6" x 10" x 8"	L.F.	14.42
3520	Sills & Stools 6"	"	32.75
3530	Splash Blocks 3" x 16"	EA.	78.75
3600	MISCELLANEOUS ADDITIONS TO ABOVE CONCRETE		
3610	Abrasives - Carborundum - Grits	S.F.	1.11
3620	Strips	L.F.	3.68
3630	Bushhammer - Green Concrete	S.F.	2.46

		UNIT	COST
03999.10	**CONCRETE, Cont'd...**		
3640	Cured Concrete	S.F.	3.31
3650	Chamfers - Plastic 3/4"	L.F.	1.30
3660	Wood 3/4"	"	0.78
3670	Metal 3/4"	"	1.42
3680	Colors Dust On	S.F.	1.03
3690	Integral (Top 1")	"	1.75
3700	Control Joints - Asphalt 1/2" x 4"	L.F.	1.09
3710	1/2" x 6"	"	1.26
3720	PolyFoam 1/2" x 4"	"	1.37
3730	Dovetail Slots 22 Ga	"	1.91
3740	24 Ga	"	1.23
3800	Hardeners Acrylic and Urethane	S.F.	0.36
3810	Epoxy	"	0.46
3820	Joint Sealers Epoxy	L.F.	4.40
3830	Rubber Asphalt	"	2.30
3840	Moisture Proofing - Polyethylene 4 mil	S.F.	0.24
3850	6 mil	"	0.28
3900	Non-Shrink Grouts - Non - Metallic	C.F.	51.00
3910	Aluminum Oxide	"	37.00
3920	Iron Oxide	"	54.25
4000	Reglets Flashing	L.F.	3.39
4010	Sand Blast - Light	S.F.	1.75
4020	Heavy	"	3.40
4030	Shelf Angle Inserts 5/8"	L.F.	12.32
4040	Stair Nosings Steel - Galvanized	"	8.88
4050	Tongue & Groove Joint Forms - Asphalt 5 1/2"	"	2.66
4060	Wood 5 1/2"	"	2.18
4070	Metal 5 1/2"	"	2.76
4100	Treads - Extruded Aluminum	"	11.69
4110	Cast Iron	"	13.52
4120	Water Stops Center Bulb - Rubber - 6"	"	13.79
4130	9"	"	26.00
4140	Polyethylene - 6"	"	3.80
4150	9"	"	3.84

Design & Construction Resources

TABLE OF CONTENTS PAGE

		UNIT	LABOR	MAT.	TOTAL
04100.10	**MASONRY GROUT**				
0100	Grout, non shrink, non-metallic, trowelable	C.F.	1.48	4.95	6.43
2110	Grout door frame, hollow metal				
2120	Single	EA.	56.00	12.25	68.25
2140	Double	"	59.00	17.25	76.25
2980	Grout-filled concrete block (CMU)				
3000	4" wide	S.F.	1.85	0.33	2.18
3020	6" wide	"	2.02	0.86	2.88
3040	8" wide	"	2.23	1.27	3.50
3060	12" wide	"	2.34	2.09	4.43
3070	Grout-filled individual CMU cells				
3090	4" wide	L.F.	1.11	0.27	1.38
3100	6" wide	"	1.11	0.37	1.48
3120	8" wide	"	1.11	0.49	1.60
3140	10" wide	"	1.27	0.61	1.88
3160	12" wide	"	1.27	0.74	2.01
4000	Bond beams or lintels, 8" deep				
4010	6" thick	L.F.	1.83	0.74	2.57
4020	8" thick	"	2.01	0.99	3.00
4040	10" thick	"	2.24	1.24	3.48
4060	12" thick	"	2.52	1.48	4.00
5000	Cavity walls				
5020	2" thick	S.F.	2.69	0.82	3.51
5040	3" thick	"	2.69	1.24	3.93
5060	4" thick	"	2.88	1.65	4.53
5080	6" thick	"	3.36	2.47	5.83
04150.10	**MASONRY ACCESSORIES**				
0200	Foundation vents	EA.	20.25	32.25	52.50
1010	Bar reinforcing				
1015	Horizontal				
1020	#3 - #4	Lb.	2.02	0.94	2.96
1030	#5 - #6	"	1.68	0.94	2.62
1035	Vertical				
1040	#3 - #4	Lb.	2.52	0.94	3.46
1050	#5 - #6	"	2.02	0.94	2.96
1100	Horizontal joint reinforcing				
1105	Truss type				
1110	4" wide, 6" wall	L.F.	0.20	0.19	0.39
1120	6" wide, 8" wall	"	0.21	0.19	0.40
1130	8" wide, 10" wall	"	0.21	0.24	0.45
1140	10" wide, 12" wall	"	0.22	0.24	0.46
1150	12" wide, 14" wall	"	0.24	0.29	0.53
1155	Ladder type				
1160	4" wide, 6" wall	L.F.	0.20	0.14	0.34
1170	6" wide, 8" wall	"	0.21	0.16	0.37
1180	8" wide, 10" wall	"	0.21	0.17	0.38
1190	10" wide, 12" wall	"	0.21	0.20	0.41
2000	Rectangular wall ties				
2005	3/16" dia., galvanized				
2010	2" x 6"	EA.	0.84	0.35	1.19
2020	2" x 8"	"	0.84	0.37	1.21
2040	2" x 10"	"	0.84	0.42	1.26

		UNIT	LABOR	MAT.	TOTAL
04150.10	**MASONRY ACCESSORIES, Cont'd...**				
2050	2" x 12"	EA.	0.84	0.48	1.32
2060	4" x 6"	"	1.01	0.39	1.40
2070	4" x 8"	"	1.01	0.45	1.46
2080	4" x 10"	"	1.01	0.58	1.59
2090	4" x 12"	"	1.01	0.67	1.68
2095	1/4" dia., galvanized				
2100	2" x 6"	EA.	0.84	0.64	1.48
2110	2" x 8"	"	0.84	0.72	1.56
2120	2" x 10"	"	0.84	0.82	1.66
2130	2" x 12"	"	0.84	0.94	1.78
2140	4" x 6"	"	1.01	0.74	1.75
2150	4" x 8"	"	1.01	0.82	1.83
2160	4" x 10"	"	1.01	0.94	1.95
2170	4" x 12"	"	1.01	0.99	2.00
2200	"Z" type wall ties, galvanized				
2215	6" long				
2220	1/8" dia.	EA.	0.84	0.30	1.14
2230	3/16" dia.	"	0.84	0.33	1.17
2240	1/4" dia.	"	0.84	0.35	1.19
2245	8" long				
2250	1/8" dia.	EA.	0.84	0.33	1.17
2260	3/16" dia.	"	0.84	0.35	1.19
2270	1/4" dia.	"	0.84	0.37	1.21
2275	10" long				
2280	1/8" dia.	EA.	0.84	0.35	1.19
2290	3/16" dia.	"	0.84	0.39	1.23
2300	1/4" dia.	"	0.84	0.45	1.29
3000	Dovetail anchor slots				
3015	Galvanized steel, filled				
3020	24 ga.	L.F.	1.26	0.82	2.08
3040	20 ga.	"	1.26	1.03	2.29
3060	16 oz. copper, foam filled	"	1.26	2.03	3.29
3100	Dovetail anchors				
3115	16 ga.				
3120	3-1/2" long	EA.	0.84	0.25	1.09
3140	5-1/2" long	"	0.84	0.30	1.14
3150	12 ga.				
3160	3-1/2" long	EA.	0.84	0.33	1.17
3180	5-1/2" long	"	0.84	0.55	1.39
3200	Dovetail, triangular galvanized ties, 12 ga.				
3220	3" x 3"	EA.	0.84	0.56	1.40
3240	5" x 5"	"	0.84	0.60	1.44
3260	7" x 7"	"	0.84	0.68	1.52
3280	7" x 9"	"	0.84	0.72	1.56
3400	Brick anchors				
3420	Corrugated, 3-1/2" long				
3440	16 ga.	EA.	0.84	0.22	1.06
3460	12 ga.	"	0.84	0.38	1.22
3500	Non-corrugated, 3-1/2" long				
3520	16 ga.	EA.	0.84	0.30	1.14
3540	12 ga.	"	0.84	0.55	1.39
3580	Cavity wall anchors, corrugated, galvanized				

		UNIT	LABOR	MAT.	TOTAL
04150.10	**MASONRY ACCESSORIES, Cont'd...**				
3600	5" long				
3620	16 ga.	EA.	0.84	0.68	1.52
3640	12 ga.	"	0.84	1.01	1.85
3660	7" long				
3680	28 ga.	EA.	0.84	0.74	1.58
3700	24 ga.	"	0.84	0.94	1.78
3720	22 ga.	"	0.84	0.96	1.80
3740	16 ga.	"	0.84	1.08	1.92
3800	Mesh ties, 16 ga., 3" wide				
3820	8" long	EA.	0.84	0.91	1.75
3840	12" long	"	0.84	1.01	1.85
3860	20" long	"	0.84	1.39	2.23
3900	24" long	"	0.84	1.54	2.38
04150.20	**MASONRY CONTROL JOINTS**				
1000	Control joint, cross shaped PVC	L.F.	1.26	2.17	3.43
1010	Closed cell joint filler				
1020	1/2"	L.F.	1.26	0.38	1.64
1040	3/4"	"	1.26	0.77	2.03
1070	Rubber, for				
1080	4" wall	L.F.	1.26	2.50	3.76
1090	6" wall	"	1.32	3.10	4.42
1100	8" wall	"	1.40	3.74	5.14
1110	PVC, for				
1120	4" wall	L.F.	1.26	1.30	2.56
1140	6" wall	"	1.32	2.20	3.52
1160	8" wall	"	1.40	3.33	4.73
04150.50	**MASONRY FLASHING**				
0080	Through-wall flashing				
1000	5 oz. coated copper	S.F.	4.21	3.35	7.56
1020	0.030" elastomeric	"	3.36	1.10	4.46
04210.10	**BRICK MASONRY**				
0100	Standard size brick, running bond				
1000	Face brick, red (6.4/sf)				
1020	Veneer	S.F.	8.42	5.39	13.81
1030	Cavity wall	"	7.21	5.39	12.60
1040	9" solid wall	"	14.50	10.75	25.25
1200	Common brick (6.4/sf)				
1210	Select common for veneers	S.F.	8.42	3.74	12.16
1215	Back-up				
1220	4" thick	S.F.	6.31	3.41	9.72
1230	8" thick	"	10.00	6.82	16.82
1235	Firewall				
1240	12" thick	S.F.	16.75	10.25	27.00
1250	16" thick	"	23.00	13.75	36.75
1300	Glazed brick (7.4/sf)				
1310	Veneer	S.F.	9.18	12.00	21.18
1400	Buff or gray face brick (6.4/sf)				
1410	Veneer	S.F.	8.42	5.98	14.40
1420	Cavity wall	"	7.21	5.98	13.19
1500	Jumbo or oversize brick (3/sf)				
1510	4" veneer	S.F.	5.05	4.12	9.17

		UNIT	LABOR	MAT.	TOTAL
04210.10	**BRICK MASONRY, Cont'd...**				
1530	4" back-up	S.F.	4.21	4.12	8.33
1540	8" back-up	"	7.21	4.78	11.99
1550	12" firewall	"	12.75	6.43	19.18
1560	16" firewall	"	16.75	9.07	25.82
1600	Norman brick, red face, (4.5/sf)				
1620	4" veneer	S.F.	6.31	5.94	12.25
1640	Cavity wall	"	5.61	5.94	11.55
3000	Chimney, standard brick, including flue				
3020	16" x 16"	L.F.	51.00	37.75	88.75
3040	16" x 20"	"	51.00	43.00	94.00
3060	16" x 24"	"	51.00	62.00	113
3080	20" x 20"	"	63.00	62.00	125
3100	20" x 24"	"	63.00	65.00	128
3120	20" x 32"	"	72.00	74.00	146
4000	Window sill, face brick on edge	"	12.75	2.97	15.72
04210.20	**STRUCTURAL TILE**				
5000	Structural glazed tile				
5010	6T series, 5-1/2" x 12"				
5020	Glazed on one side				
5040	2" thick	S.F.	5.05	10.25	15.30
5060	4" thick	"	5.05	12.00	17.05
5080	6" thick	"	5.61	19.00	24.61
5100	8" thick	"	6.31	23.25	29.56
5200	Glazed on two sides				
5220	4" thick	S.F.	6.31	20.00	26.31
5240	6" thick	"	7.21	24.25	31.46
5500	Special shapes				
5510	Group 1	S.F.	10.00	11.25	21.25
5520	Group 2	"	10.00	14.25	24.25
5530	Group 3	"	10.00	19.00	29.00
5540	Group 4	"	10.00	38.00	48.00
5550	Group 5	"	10.00	46.50	56.50
5600	Fire rated				
5620	4" thick, 1 hr rating	S.F.	5.05	20.00	25.05
5640	6" thick, 2 hr rating	"	5.61	24.50	30.11
6000	8W series, 8" x 16"				
6010	Glazed on one side				
6020	2" thick	S.F.	3.36	11.50	14.86
6040	4" thick	"	3.36	12.25	15.61
6060	6" thick	"	3.88	20.00	23.88
6080	8" thick	"	3.88	22.00	25.88
6100	Glazed on two sides				
6120	4" thick	S.F.	4.21	17.25	21.46
6140	6" thick	"	5.05	24.25	29.30
6160	8" thick	"	5.05	29.50	34.55
6200	Special shapes				
6220	Group 1	S.F.	7.21	15.50	22.71
6230	Group 2	"	7.21	19.00	26.21
6240	Group 3	"	7.21	20.75	27.96
6250	Group 4	"	7.21	34.50	41.71
6260	Group 5	"	7.21	75.00	82.21

		UNIT	LABOR	MAT.	TOTAL
04210.20	**STRUCTURAL TILE, Cont'd...**				
6270	Fire rated				
6290	4" thick, 1 hr rating	S.F.	7.21	26.00	33.21
6300	6" thick, 2 hr rating	"	7.21	31.25	38.46
04210.60	**PAVERS, MASONRY**				
4000	Brick walk laid on sand, sand joints				
4020	Laid flat, (4.5 per sf)	S.F.	5.61	3.78	9.39
4040	Laid on edge, (7.2 per sf)	"	8.42	6.05	14.47
5000	Precast concrete patio blocks				
5005	2" thick				
5010	Natural	S.F.	1.68	2.68	4.36
5020	Colors	"	1.68	3.74	5.42
5080	Exposed aggregates, local aggregate				
5100	Natural	S.F.	1.68	6.16	7.84
5120	Colors	"	1.68	6.71	8.39
5130	Granite or limestone aggregate	"	1.68	8.44	10.12
5140	White tumblestone aggregate	"	1.68	5.17	6.85
6000	Stone pavers, set in mortar				
6005	Bluestone				
6008	1" thick				
6010	Irregular	S.F.	12.75	6.14	18.89
6020	Snapped rectangular	"	10.00	9.35	19.35
6060	1-1/2" thick, random rectangular	"	12.75	10.75	23.50
6070	2" thick, random rectangular	"	14.50	12.75	27.25
6090	Slate				
6100	Natural cleft				
6110	Irregular, 3/4" thick	S.F.	14.50	6.71	21.21
6115	Random rectangular				
6120	1-1/4" thick	S.F.	12.75	14.50	27.25
6130	1-1/2" thick	"	14.00	16.25	30.25
7000	Granite blocks				
7010	3" thick, 3" to 6" wide				
7020	4" to 12" long	S.F.	16.75	13.75	30.50
7030	6" to 15" long	"	14.50	11.00	25.50
9000	Flagstone pavers				
9010	Random sizes, 1-4 s.f., tumbled cobble	S.F.	10.00	15.50	25.50
9020	Tumbled patio	"	10.00	16.50	26.50
9040	Saw-cut Flagstone Tiles				
9060	12"x12", assorted colors	S.F.	7.21	11.00	18.21
9080	18"x18", assorted colors	"	6.31	11.00	17.31
9100	24"x24", assorted colors	"	5.61	11.00	16.61
9800	Crushed stone, white marble, 3" thick	"	0.81	1.70	2.51
04220.10	**CONCRETE MASONRY UNITS**				
0110	Hollow, load bearing				
0120	4"	S.F.	3.74	1.37	5.11
0140	6"	"	3.88	2.01	5.89
0160	8"	"	4.21	2.31	6.52
0180	10"	"	4.59	3.19	7.78
0190	12"	"	5.05	3.66	8.71
0280	Solid, load bearing				
0300	4"	S.F.	3.74	2.15	5.89
0320	6"	"	3.88	2.42	6.30

04220.10	CONCRETE MASONRY UNITS, Cont'd...	UNIT	LABOR	MAT.	TOTAL
0340	8"	S.F.	4.21	3.30	7.51
0360	10"	"	4.59	3.52	8.11
0380	12"	"	5.05	5.22	10.27
0480	Back-up block, 8" x 16"				
0500	2"	S.F.	2.88	1.32	4.20
0540	4"	"	2.97	1.37	4.34
0560	6"	"	3.15	2.01	5.16
0580	8"	"	3.36	2.31	5.67
0600	10"	"	3.60	3.19	6.79
0620	12"	"	3.88	3.66	7.54
0980	Foundation wall, 8" x 16"				
1000	6"	S.F.	3.60	2.01	5.61
1030	8"	"	3.88	2.31	6.19
1040	10"	"	4.21	3.19	7.40
1050	12"	"	4.59	3.66	8.25
1055	Solid				
1060	6"	S.F.	3.88	2.42	6.30
1070	8"	"	4.21	3.30	7.51
1080	10"	"	4.59	3.52	8.11
1100	12"	"	5.05	5.22	10.27
1480	Exterior, styrofoam inserts, standard weight, 8" x 16"				
1500	6"	S.F.	3.88	3.94	7.82
1530	8"	"	4.21	4.24	8.45
1540	10"	"	4.59	5.44	10.03
1550	12"	"	5.05	7.60	12.65
1580	Lightweight				
1600	6"	S.F.	3.88	4.41	8.29
1660	8"	"	4.21	4.96	9.17
1680	10"	"	4.59	5.28	9.87
1700	12"	"	5.05	6.96	12.01
1980	Acoustical slotted block				
2000	4"	S.F.	4.59	4.75	9.34
2020	6"	"	4.59	4.98	9.57
2040	8"	"	5.05	6.21	11.26
2050	Filled cavities				
2060	4"	S.F.	5.61	5.09	10.70
2070	6"	"	5.94	5.87	11.81
2080	8"	"	6.31	7.52	13.83
4000	Hollow, split face				
4020	4"	S.F.	3.74	3.25	6.99
4030	6"	"	3.88	3.77	7.65
4040	8"	"	4.21	3.96	8.17
4080	10"	"	4.59	4.43	9.02
4100	12"	"	5.05	4.73	9.78
4480	Split rib profile				
4500	4"	S.F.	4.59	3.96	8.55
4520	6"	"	4.59	4.59	9.18
4540	8"	"	5.05	5.00	10.05
4560	10"	"	5.05	5.47	10.52
4580	12"	"	5.05	5.94	10.99
4980	High strength block, 3500 psi				
5000	2"	S.F.	3.74	1.54	5.28

04220.10	CONCRETE MASONRY UNITS, Cont'd...	UNIT	LABOR	MAT.	TOTAL
5020	4"	S.F.	3.88	1.92	5.80
5030	6"	"	3.88	2.29	6.17
5040	8"	"	4.21	2.59	6.80
5050	10"	"	4.59	3.02	7.61
5060	12"	"	5.05	3.58	8.63
5500	Solar screen concrete block				
5505	4" thick				
5510	6" x 6"	S.F.	11.25	4.58	15.83
5520	8" x 8"	"	10.00	5.47	15.47
5530	12" x 12"	"	7.77	5.61	13.38
5540	8" thick				
5550	8" x 16"	S.F.	7.21	5.61	12.82
7000	Glazed block				
7020	Cove base, glazed 1 side, 2"	L.F.	5.61	10.50	16.11
7030	4"	"	5.61	10.75	16.36
7040	6"	"	6.31	11.00	17.31
7050	8"	"	6.31	11.75	18.06
7055	Single face				
7060	2"	S.F.	4.21	11.00	15.21
7080	4"	"	4.21	13.50	17.71
7090	6"	"	4.59	14.50	19.09
7100	8"	"	5.05	15.25	20.30
7105	10"	"	5.61	17.25	22.86
7110	12"	"	5.94	18.25	24.19
7115	Double face				
7120	4"	S.F.	5.31	16.25	21.56
7140	6"	"	5.61	19.25	24.86
7160	8"	"	6.31	20.25	26.56
7180	Corner or bullnose				
7200	2"	EA.	6.31	11.00	17.31
7240	4"	"	7.21	14.00	21.21
7260	6"	"	7.21	17.25	24.46
7280	8"	"	8.42	18.75	27.17
7290	10"	"	9.18	20.25	29.43
7300	12"	"	10.00	22.00	32.00
9500	Gypsum unit masonry				
9510	Partition blocks (12"x30")				
9515	Solid				
9520	2"	S.F.	2.02	1.11	3.13
9525	Hollow				
9530	3"	S.F.	2.02	1.12	3.14
9540	4"	"	2.10	1.28	3.38
9550	6"	"	2.29	1.37	3.66
9900	Vertical reinforcing				
9920	4' o.c., add 5% to labor				
9940	2'8" o.c., add 15% to labor				
9960	Interior partitions, add 10% to labor				
04220.90	BOND BEAMS & LINTELS				
0980	Bond beam, no grout or reinforcement				
0990	8" x 16" x				
1000	4" thick	L.F.	3.88	1.39	5.27

		UNIT	LABOR	MAT.	TOTAL
04220.90	**BOND BEAMS & LINTELS, Cont'd...**				
1040	6" thick	L.F.	4.04	2.13	6.17
1060	8" thick	"	4.21	2.44	6.65
1080	10" thick	"	4.39	3.02	7.41
1100	12" thick	"	4.59	3.43	8.02
6000	Beam lintel, no grout or reinforcement				
6010	8" x 16" x				
6020	10" thick	L.F.	5.05	5.58	10.63
6040	12" thick	"	5.61	7.09	12.70
6080	Precast masonry lintel				
7000	6 lf, 8" high x				
7020	4" thick	L.F.	8.42	5.36	13.78
7040	6" thick	"	8.42	6.85	15.27
7060	8" thick	"	9.18	7.75	16.93
7080	10" thick	"	9.18	9.25	18.43
7090	10 lf, 8" high x				
7100	4" thick	L.F.	5.05	6.74	11.79
7120	6" thick	"	5.05	8.31	13.36
7140	8" thick	"	5.61	9.25	14.86
7160	10" thick	"	5.61	12.50	18.11
8000	Steel angles and plates				
8010	Minimum	Lb.	0.72	1.04	1.76
8020	Maximum	"	1.26	1.54	2.80
8200	Various size angle lintels				
8205	1/4" stock				
8210	3" x 3"	L.F.	3.15	5.39	8.54
8220	3" x 3-1/2"	"	3.15	5.94	9.09
8225	3/8" stock				
8230	3" x 4"	L.F.	3.15	9.35	12.50
8240	3-1/2" x 4"	"	3.15	9.82	12.97
8250	4" x 4"	"	3.15	10.75	13.90
8260	5" x 3-1/2"	"	3.15	11.50	14.65
8262	6" x 3-1/2"	"	3.15	12.75	15.90
8265	1/2" stock				
8280	6" x 4"	L.F.	3.15	14.25	17.40
04240.10	**CLAY TILE**				
0100	Hollow clay tile, for back-up, 12" x 12"				
1000	Scored face				
1010	Load bearing				
1020	4" thick	S.F.	3.60	6.56	10.16
1040	6" thick	"	3.74	7.65	11.39
1060	8" thick	"	3.88	9.53	13.41
1080	10" thick	"	4.04	11.75	15.79
1100	12" thick	"	4.21	20.00	24.21
2000	Non-load bearing				
2020	3" thick	S.F.	3.48	12.75	16.23
2040	4" thick	"	3.60	6.16	9.76
2060	6" thick	"	3.74	7.12	10.86
2080	8" thick	"	3.88	9.08	12.96
2100	12" thick	"	4.21	13.00	17.21
4100	Partition, 12" x 12"				
4150	In walls				

		UNIT	LABOR	MAT.	TOTAL
04240.10	**CLAY TILE, Cont'd...**				
4201	3" thick	S.F.	4.21	5.32	9.53
4210	4" thick	"	4.21	6.17	10.38
4220	6" thick	"	4.39	6.79	11.18
4230	8" thick	"	4.59	8.91	13.50
4240	10" thick	"	4.81	10.50	15.31
4250	12" thick	"	5.05	15.50	20.55
4300	Clay tile floors				
4320	4" thick	S.F.	2.80	6.16	8.96
4330	6" thick	"	2.97	7.65	10.62
4340	8" thick	"	3.15	9.53	12.68
4350	10" thick	"	3.36	11.75	15.11
4360	12" thick	"	3.60	17.50	21.10
6000	Terra cotta				
6020	Coping, 10" or 12" wide, 3" thick	L.F.	10.00	14.75	24.75
04270.10	**GLASS BLOCK**				
1000	Glass block, 4" thick				
1040	6" x 6"	S.F.	16.75	26.00	42.75
1060	8" x 8"	"	12.75	16.50	29.25
1080	12" x 12"	"	10.00	21.00	31.00
8980	Replacement glass blocks, 4" x 8" x 8"				
9100	Minimum	S.F.	51.00	20.00	71.00
9120	Maximum	"	100	25.50	126
04295.10	**PARGING / MASONRY PLASTER**				
0080	Parging				
0100	1/2" thick	S.F.	3.36	0.29	3.65
0200	3/4" thick	"	4.21	0.37	4.58
0300	1" thick	"	5.05	0.50	5.55
04400.10	**STONE**				
0160	Rubble stone				
0180	Walls set in mortar				
0200	8" thick	S.F.	12.75	15.00	27.75
0220	12" thick	"	20.25	18.25	38.50
0420	18" thick	"	25.25	24.00	49.25
0440	24" thick	"	33.75	30.25	64.00
0445	Dry set wall				
0450	8" thick	S.F.	8.42	17.00	25.42
0455	12" thick	"	12.75	19.00	31.75
0460	18" thick	"	16.75	26.50	43.25
0465	24" thick	"	20.25	32.25	52.50
0480	Cut stone				
0490	Imported marble				
0510	Facing panels				
0520	3/4" thick	S.F.	20.25	33.25	53.50
0530	1-1/2" thick	"	23.00	52.00	75.00
0540	2-1/4" thick	"	28.00	63.00	91.00
0600	Base				
0610	1" thick				
0620	4" high	L.F.	25.25	16.50	41.75
0640	6" high	"	25.25	20.00	45.25
0700	Columns, solid				
0720	Plain faced	C.F.	340	120	460

		UNIT	LABOR	MAT.	TOTAL
04400.10	**STONE, Cont'd...**				
0740	Fluted	C.F.	340	320	660
0780	Flooring, travertine, minimum	S.F.	7.77	16.00	23.77
0800	Average	"	10.00	21.50	31.50
0820	Maximum	"	11.25	38.50	49.75
1000	Domestic marble				
1020	Facing panels				
1040	7/8" thick	S.F.	20.25	31.75	52.00
1060	1-1/2" thick	"	23.00	47.75	70.75
1080	2-1/4" thick	"	28.00	58.00	86.00
1500	Stairs				
1510	12" treads	L.F.	25.25	28.50	53.75
1520	6" risers	"	16.75	21.00	37.75
1525	Thresholds, 7/8" thick, 3' long, 4" to 6" wide				
1530	Plain	EA.	42.00	27.00	69.00
1540	Beveled	"	42.00	30.00	72.00
1545	Window sill				
1550	6" wide, 2" thick	L.F.	20.25	15.25	35.50
1555	Stools				
1560	5" wide, 7/8" thick	L.F.	20.25	22.25	42.50
1620	Limestone panels up to 12' x 5', smooth finish				
1630	2" thick	S.F.	8.92	29.75	38.67
1650	3" thick	"	8.92	34.75	43.67
1660	4" thick	"	8.92	49.75	58.67
1760	Miscellaneous limestone items				
1770	Steps, 14" wide, 6" deep	L.F.	33.75	63.00	96.75
1780	Coping, smooth finish	C.F.	16.75	88.00	105
1790	Sills, lintels, jambs, smooth finish	"	20.25	88.00	108
1800	Granite veneer facing panels, polished				
1810	7/8" thick				
1820	Black	S.F.	20.25	46.25	66.50
1840	Gray	"	20.25	36.25	56.50
1850	Base				
1860	4" high	L.F.	10.00	19.50	29.50
1870	6" high	"	11.25	23.50	34.75
1880	Curbing, straight, 6" x 16"	"	37.25	21.50	58.75
1890	Radius curbs, radius over 5'	"	49.50	26.25	75.75
1900	Ashlar veneer				
1905	4" thick, random	S.F.	20.25	35.75	56.00
1910	Pavers, 4" x 4" split				
1915	Gray	S.F.	10.00	35.00	45.00
1920	Pink	"	10.00	34.50	44.50
1930	Black	"	10.00	34.00	44.00
2000	Slate, panels				
2010	1" thick	S.F.	20.25	25.00	45.25
2020	2" thick	"	23.00	34.00	57.00
2030	Sills or stools				
2040	1" thick				
2060	6" wide	L.F.	20.25	11.75	32.00
2080	10" wide	"	22.00	19.00	41.00
2100	2" thick				
2120	6" wide	L.F.	23.00	19.00	42.00
2140	10" wide	"	25.25	31.50	56.75

04400.10	STONE, Cont'd...	UNIT	LABOR	MAT.	TOTAL
2200	Simulated masonry				
2210	Cultured stone veneer, 1-3/4" thick				
2220	Ledgestone, large-pattern	S.F.	5.05	9.35	14.40
2230	Small-pattern	"	6.73	9.35	16.08
2240	Drystack	"	5.74	9.35	15.09
2250	Castlestone	"	5.15	9.35	14.50
2260	Fieldstone	"	5.61	9.35	14.96
2270	Limestone	"	4.59	9.35	13.94
2280	Split-face	"	5.05	9.35	14.40
2290	Stream stone (river rock)	"	5.31	9.35	14.66
0030	Flagstone, strip veneer, random lengths				
0040	1" to 2-1/2" high	S.F.	5.61	9.35	14.96
0050	2" to 4" high	"	4.59	9.35	13.94
0060	5" to 6" high	"	4.04	9.35	13.39
0070	7" to 8" high	"	3.36	9.35	12.71

		UNIT	COST
04999.10	**MASONRY**		
1000	BRICK MASONRY		
1100	Conventional (Modular Size)		
1110	Running Bond - 8" x 2 2/3" x 4"	S.F.	17.20
1120	Common Bond - 6 Course Header	"	20.09
1130	Stack Bond	"	17.54
1140	Dutch & English Bond - Every Other Course Header	"	26.00
1150	Every Course Header	"	28.75
1160	Flemish Bond - Every Other Course Header	"	19.60
1170	Every Course Header	"	23.35
1200	Add: If Scaffold Needed	"	0.87
1210	Add: For each 10' of Floor Hgt. or Floor (3%)	"	0.48
1220	Add: Piers and Corbels (15% to Labor)	"	2.48
1230	Add: Sills and Soldiers (20% to Labor)	"	2.37
1240	Add: Floor Brick (10% to Labor)	"	1.73
1300	Add: Weave & Herringbone Patterns (20% to Labor)	"	2.36
1310	Add: Stack Bond (8% to Labor)	"	1.37
1320	Add: Circular or Radius Work 20% to Labor)	"	3.61
1330	Add: Rock Faced & Slurried Face (10% to Labor)	"	1.73
1340	Add: Arches (75% to Labor)	"	5.31
1350	Add: For Winter Work (below 40°)	"	1.73
1400	Production Loss (10% to Labor)	"	1.73
1410	Enclosures - Wall Area Conventional	"	1.73
1420	Heat and Fuel - Wall Area Conventional	"	0.35
1500	Econo - 8" x 4" x 3"	"	14.30
1600	Panel - 8" x 8" x 4"	"	11.22
1700	Norman - 12" x 2 2/3" x 4"	"	14.32
1800	King Size - 10" x 2 5/8" x 4"	"	12.44
1900	Norwegian - 12" x 3 1/5" x 4"	"	11.98
2000	Saxon-Utility - 12" x 4" x 3"	"	11.24
2010	12" x 4" x 4"	"	11.29
2020	12" x 4" x 6"	"	14.25
2030	12" x 4" x 8"	"	17.11
2200	Adobe - 12" x 3" x 4"	"	19.47
2300	COATED BRICK (Ceramic)	"	23.21
2400	COMMON BRICK (Clay, Concrete and Sand Lime)	"	13.24
2500	FIRE BRICK -Light Duty	"	21.62
2510	Heavy Duty	"	27.50
2520	Deduct for Residential Work - All Above - 5%		
04999.20	**CONCRETE BLOCK**		
1000	CONVENTIONAL (Struck 2 Sides - Partitions)		
1010	12" x 8"x 16" Plain	S.F.	9.16
1020	Bond Beam (with Fill-Reinforcing)	"	12.12
1030	12" x 8"x 8" Half Block	"	10.91
1040	Double End - Header	"	9.38
1050	8" x 8" x 16" Plain	"	7.99
1060	Bond Beam (with Fill-Reinforcing)	"	9.81
1070	8" x 8" x 8" Half Block	"	7.83
1080	Double End - Header	"	8.31
1090	6 "x 8" x 16" Plain	"	7.46
2000	Bond Beam (with Fill-Reinforcing)	"	8.36
2010	Half Block	"	7.35

		UNIT	COST
04999.20	**CONCRETE BLOCK, Cont'd...**		
2020	4" x 8" x 16" Plain	S.F.	7.22
2030	Half Block	"	6.29
2040	16" x 8" x 16" Plain	"	10.38
2050	Half Block	"	9.90
2060	14" x 8" x 16" Plain	"	9.86
2070	Half Block	"	9.36
2080	10" x 8" x 16" Plain	"	8.54
3000	Bond Beam (with Fill-Reinforcing)	"	9.84
3010	Half Block	"	9.58
3020	Deduct: Block Struck or Cleaned One Side	"	0.30
3030	Deduct: Block Not Struck or Cleaned Two Sides	"	0.47
3040	Deduct: Clean One Side Only	"	0.21
3050	Deduct: Lt. Wt. Block (Labor Only)	"	0.18
4000	Add: Jamb and Sash Block	"	0.50
4010	Add: Bullnose Block	"	0.74
4020	Add: Pilaster, Pier, Pedestal Work	"	0.83
4030	Add: Stack Bond Work	"	0.31
4040	Add: Radius or Circular Work	"	1.54
5000	Add: If Scaffold Needed	"	0.85
5010	Add: Winter Production Cost (10% Labor)	"	0.50
5020	Add: Winter Enclosing - Wall Area	"	0.50
5030	Add: Winter Heating - Wall Area	"	0.45
5040	Add: Core & Beam Filling - See 0411 (Page 4A-18)		
5050	Add: Insulation - See 0412 (Page 4A-18)		
6000	Deduct for Residential Work - 5%		
8010	SCREEN WALL - 4" x 12" x 12"	S.F.	7.13
8020	BURNISHED - 12" x 8" x 16"	"	13.36
8030	8" x 8" x 16"	"	12.45
8040	6" x 8" x 16"	"	11.38
8050	4" x 8" x 16"	"	10.67
8060	2" x 8" x 16"	"	9.62
8070	Add for Shapes	"	2.89
8080	Add for 2 Faced Finish	"	4.17
8090	Add for Scored Finish - 1 Face	"	1.01
8100	PREFACED UNITS -12" x 8" x 16" Stretcher	"	18.87
8110	(Ceramic Glazed) 12" x 8" x 16" Glazed 2 Face	"	25.00
8120	8" x 8" x 16" Stretcher	"	17.34
8130	Glazed 2 Face	"	23.76
8140	6" x 8" x 16" Stretcher	"	15.61
8150	Glazed 2 Face	"	23.69
8160	4" x 8" x 16" Stretcher	"	16.38
8170	Glazed 2 Face	"	22.74
8180	2" x 8" x 16" Stretcher	"	16.25
8190	4" x 16" x 16" Stretcher	"	41.75
8200	Add for Base, Caps, Jambs, Headers, Lintels	"	4.68
8210	Add for Scored Block	"	1.38
04999.30	**CLAY BACKING AND PARTITION TILE**		
0010	3" x 12" x 12"	S.F.	6.22
0020	4" x 12" x 12"	"	6.67
0030	6" x 12" x 12"	"	7.94
0040	8" x 12" x 12"	"	8.70

		UNIT	COST
04999.40	**CLAY FACING TILE (GLAZED)**		
1000	6T or 5" x 12" - SERIES		
1010	2" x 5 1/3" x 12" Soap Stretcher (Solid Back)	S.F.	22.74
1020	4" x 5 1/2" x 12" 1 Face Stretcher	"	24.37
1030	2 Face Stretcher	"	28.75
1040	6" x 5 1/3" x 12" 1 Face Stretcher	"	29.25
1050	8" x 5 1/3" x 12" 1 Face Stretcher	"	16.70
2000	8W or 8" x 16" - SERIES		
2010	2" x 8" x 16" Soap Stretcher (Solid Back)	S.F.	17.25
2020	4" x 8" x 16" 1 Face Stretcher	"	17.72
2030	2 Face Stretcher	"	23.46
2040	6" x 8" x 16" 1 Face Stretcher	"	21.07
2050	8" x 8" x 16" 1 Face Stretcher	"	21.97
2060	Add for Shapes (Average)	PCT.	50.00
2070	Add for Designer Colors	"	20.00
2080	Add for Less than Truckload Lots	"	10.00
2090	Add for Base Only	"	25.00
04999.50	**GLASS UNITS**		
0010	4" x 8" x 4"	S.F.	57.50
0020	6" x 6" x 4"	"	61.50
0030	8" x 8" x 4"	"	37.25
0040	12" x 12" x 4"	"	37.75
04999.60	**TERRA-COTTA**		
0010	Unglazed	S.F.	12.96
0020	Glazed	"	17.34
0030	Colored Glazed	"	21.07
04999.70	**FLUE LINING**		
0010	8" x 12"	S.F.	21.59
0020	12" x 12"	"	23.37
0030	16" x 16"	"	37.25
0040	18" x 18"	"	38.00
0050	20" x 20"	"	75.00
0060	24" x 24"	"	95.50
04999.80	**NATURAL STONE**		
1000	CUT STONE BY S.F.		
1010	Limestone- Indiana and Alabama - 3"	S.F.	43.00
1020	4"	"	47.50
1030	Minnesota, Wisc, Texas, etc.- 3"	"	48.25
1040	4"	"	55.00
1200	Marble - 2"	"	59.50
1210	3"	"	63.50
1300	Granite - 2"	"	55.50
1310	3"	"	60.00
1400	Slate - 1 1/2"	"	59.50
2000	Ashlar - 4" Sawed Bed		
2100	Limestone- Indiana - Random	S.F.	33.75
2110	Coursed - 2" - 5" - 8"	"	32.25
2120	Minnesota, Alabama, Wisconsin, etc.		
2130	Split Face - Random	S.F.	33.75
2140	Coursed	"	38.00
2150	Sawed or Planed Face - Random	"	36.50
2160	Coursed	"	38.50

		UNIT	COST
04999.80	**NATURAL STONE, Cont'd...**		
2200	Marble - Sawed - Random	S.F.	48.00
2210	Coursed	"	48.75
2300	Granite - Bushhammered - Random	"	49.00
2310	Coursed	"	54.00
2400	Quartzite	"	37.50
3000	ROUGH STONE		
3100	Rubble and Flagstone	S.F.	34.50
3200	Field Stone or Boulders	"	32.25
3300	Light Weight Boulders (Igneous)		
3310	2" to 4" Veneer - Sawed Back	S.F.	28.00
3320	3" to 10" Boulders	"	29.50
3330	CUT STONE BY C.F		
3400	Limestone- Indiana and Alabama - 3"	C.F.	187
3410	4"	"	136
3420	Minnesota, Wisc, Texas, etc.- 3"	"	196
3430	4"	"	184
3500	Marble - 2"	"	410
3510	3"	"	260
3600	Granite - 2"	"	380
3610	3"	"	260
3700	Slate - 1 1/2"	"	390
3800	ASHLAR Limestone- Indiana - Random	"	100
3810	Coursed - 2" - 5" - 8"	"	99.75
3820	Split Face - Random	"	101
3830	Coursed	"	124
3840	Sawed or Planed Face - Random	"	123
3850	Coursed	"	125
3900	Marble - Sawed - Random	"	153
3910	Coursed	"	154
4000	Granite - Bushhammered - Random	"	156
4010	Coursed	"	160
4100	Quartzite	"	123
4200	ROUGH STONE, Rubble and Flagstone	"	101
4300	Field Stone or Boulders	"	99.75
04999.90	**PRECAST VENEERS AND SIMULATED MASONRY**		
1000	ARCHITECTURAL PRECAST STONE		
1005	Limestone	S.F.	42.25
1010	Marble	"	50.50
2010	PRECAST CONCRETE	"	34.50
3010	MOSAIC GRANITE PANELS	"	47.75
04999.91	**MORTAR**		
0010	Portland Cement and Lime Mortar - Labor in Unit Costs	S.F.	126
0020	Masonry Cement Mortar - Labor in Unit Costs	"	121
0030	Mortar for Standard Brick - Material Only	"	0.52
04999.92	**CORE FILLING FOR REINFORCED CONCRETE**		
1010	Add to Block Prices in 0402.0 (Job Mixed)		
1020	12" x 8" x 16" Plain (includes #4 Rod 16" O.C. Vertical)	S.F.	3.95
1030	Bond Beam (includes 2 #4 Rods)	"	4.65
1040	8" x 8" x 6" Plain (includes #4 Rod 16" O.C. Vertical)	"	2.40
1050	Bond Beam (includes 2 #4 Rods)	"	1.96
1060	6" x 8" x 16" Plain	"	2.13

		UNIT	COST
04999.92	**CORE FILLING FOR REINFORCED CONCRETE BLOCK, Cont'd...**		
1070	Bond Beam (includes 1 #4 Rod)	S.F.	1.09
1080	10" x 8" x 16" Bond Beam (includes 2 #4 Rods)	"	4.22
1090	14" x 8" x 16" Bond Beam (includes 2 #4 Rods)	"	4.82
04999.93	**CORE AND CAVITY FILL FOR INSULATED**		
2100	LOOSE		
2110	Core Fill - Expanded Styrene		
2120	12" x 8" x 16" Concrete Block	S.F.	1.49
2130	10" x 8" x 16" Concrete Block	"	1.32
2140	8" x 8" x 16" Concrete Block	"	0.94
2150	6" x 8" x 16" Concrete Block	"	0.81
2160	8" Brick - Jumbo - Thru the Wall	"	0.71
2170	6" Brick - Jumbo - Thru the Wall	"	0.57
2180	4" Brick - Jumbo	"	0.49
2190	Cavity Fill - per Inch		
2200	Expanded Styrene	S.F.	0.64
2210	Mica	"	1.06
2220	Fiber Glass	"	0.65
2230	Rock Wool	"	0.65
2240	Cellulose	"	0.64
2300	RIGID - Fiber Glass - 3# Density, 1"	"	1.10
2310	2"1.58		
2320	Expanded Styrene - Molded - 1# Density, 1"	S.F.	0.81
2330	2"	"	1.20
2340	Extruded - 2# Density, 1"	"	1.16
2350	2"	"	1.76
2360	Expanded Urethane - 1"	"	1.25
2370	2"	"	1.65
2380	Perlite - 1"	"	1.20
2390	2"	"	1.87
2400	Add for Embedded Water Type	"	1.87
2410	Add for Glued Applications	"	0.07
04999.95	**CLEANING AND POINTING**		
4010	Brick and Stone - Acid or Chemicals	S.F.	0.89
4020	Soap and Water	"	0.84
4030	Block & Facing Tile - 1 Face (Incl. Hollow Metal Frames)	"	0.46
4040	Point with White Cement	"	1.77
04999.97	**SCAFFOLD**		
7100	EQUIPMENT, TOOLS AND BLADES		
7110	Percentage of Labor as an average	PCT.	7.00
7120	See Division 1 for Rental Rates & New Costs		
7200	SCAFFOLD- Tubular Frame - to 40 feet - Exterior	S.F.	1.11
7210	Tubular Frame - to 16 feet - Interior	"	1.06
7220	Swing Stage - 40 feet and up	"	1.17

Design & Construction Resources

TABLE OF CONTENTS   PAGE

		UNIT	LABOR	MAT.	TOTAL
05050.10	**STRUCTURAL WELDING**				
0080	Welding				
0100	Single pass				
0120	1/8"	L.F.	2.90	0.33	3.23
0140	3/16"	"	3.87	0.55	4.42
0160	1/4"	"	4.84	0.77	5.61
05050.30	**MECHANICAL ANCHORS**				
0010	Drop-in, stainless steel, for masonry, 3/8"	EA.			0.97
0020	1/2"	"			1.07
0030	Hollow wall, for use in gypsum drywall, 6-32"x				
0040	1-1/4"	EA.			0.33
0050	2-3/8"	"			0.38
0060	1-3/4"	"			0.35
0070	1-1/2"	"			0.33
0080	2"	"			0.35
1000	Concrete anchor, 3/16"x				
1010	1-1/4"	EA.			0.44
1020	1-3/4"	"			0.55
1030	2-3/4"	"			0.55
1040	1/4"x				
1050	1-1/4"	EA.			0.55
1060	2-1/4"	"			0.66
1070	2-3/4"	"			0.71
1080	Toggle anchor, 5/8"	"			1.50
2000	3/4"	"			1.50
2010	Toggle bolts, 1/8"x				
2020	2"	EA.			0.27
2030	3"	"			0.27
2040	4"	"			0.27
2050	3/16"x				
2060	2"	EA.			0.55
2070	3"	"			0.55
2080	4"	"			0.55
2090	1/4"x				
3000	2"	EA.			0.80
3010	3"	"			0.80
3020	4"	"			0.80
3030	6"	"			1.61
3050	Sleeve anchor, 3/8"x				
3060	1-7/8"	EA.			0.34
3070	2-1/4"	"			0.38
3080	3"	"			0.44
3090	1/4x2-1/4"	"			0.39
4000	1/2"x				
4010	1/4"	EA.			0.44
4020	3"	"			0.49
4030	4"	"			0.66
4040	5/8"x				
4050	1/4"	EA.			0.77
4060	4-1/4"	"			0.77
4070	6"	"			0.82
4080	Machine screw, corrosion resistant, for use in masonry, 5/8"x				

		UNIT	LABOR	MAT.	TOTAL
05050.30	**MECHANICAL ANCHORS, Cont'd...**				
4090	2"	EA.			0.72
5000	4"	"			1.10
5010	6"	"			1.85
5020	8"	"			2.69
5030	Wedge anchor, 1/4"x2-1/4"	"			0.38
5040	3/8"x				
5050	2-1/4"	EA.			0.38
5060	3"	"			0.49
5070	3/4"	"			0.55
5080	5"	"			0.66
5090	1/2"x				
6000	3/4"	EA.			0.52
6010	3-1/4"	"			0.96
6020	4-1/4"	"			1.15
6030	5-1/2"	"			1.43
6040	5/8"x				
6050	6"	EA.			2.05
6060	7"	"			2.40
6070	8"	"			2.77
6080	Spring wing, toggle bolt, for use in hollow walls, 1/8"x				
6090	2"	EA.			0.39
7000	3"	"			0.39
7010	4"	"			0.39
7020	3/16"x				
7030	2"	EA.			0.55
7040	3"	"			0.55
7050	4"	"			0.55
7060	5"	"			0.55
7070	1/4"x				
7080	2"	EA.			0.80
7090	3"	"			0.80
8000	4"	"			0.80
8010	3/8"x6"	"			1.61
8020	Stud anchor, 1-1/4"	"			0.55
8030	3/4"	"			0.55
8040	Hex bolt, zinc plated, 3/4"x				
8050	4"	EA.			2.47
8060	6"	"			3.05
8070	8"	"			3.63
8080	1/2"x				
8090	4"	EA.			0.85
9000	6"	"			1.33
9010	8"	"			1.69
9020	Lag screw, 1/4"x				
9030	2"	EA.			0.17
9040	4"	"			0.38
9050	6"	"			0.61
9060	Machine screw, zinc plated, 5/8"x				
9070	4"	EA.			1.10
9100	6"	"			1.85
9110	8"	"			2.69
9120	Ribbed plastic anchor 5/8"x				

		UNIT	LABOR	MAT.	TOTAL
05050.30	**MECHANICAL ANCHORS, Cont'd...**				
9130	1-1/4"	EA.			0.23
9140	1-1/2"	"			0.33
05050.90	**METAL ANCHORS**				
1000	Anchor bolts				
1020	3/8" x				
1040	8" long	EA.			0.97
1080	12" long	"			1.16
1090	1/2" x				
1100	8" long	EA.			1.46
1140	12" long	"			1.70
1170	5/8" x				
1180	8" long	EA.			1.36
1220	12" long	"			1.60
1270	3/4" x				
1280	8" long	EA.			1.94
1300	12" long	"			2.18
4480	Non-drilling anchor				
4500	1/4"	EA.			0.64
4540	3/8"	"			0.80
4560	1/2"	"			1.23
7000	Self-drilling anchor				
7020	1/4"	EA.			1.62
7060	3/8"	"			2.44
7080	1/2"	"			3.25
05050.95	**METAL LINTELS**				
0080	Lintels, steel				
0100	Plain	Lb.	1.45	1.33	2.78
0120	Galvanized	"	1.45	2.00	3.45
05120.10	**STRUCTURAL STEEL**				
0100	Beams and girders, A-36				
0120	Welded	TON	670	2,650	3,320
0140	Bolted	"	610	2,590	3,200
0180	Columns				
0185	Pipe				
0190	6" dia.	Lb.	0.66	1.49	2.15
1300	Structural tube				
1310	6" square				
1320	Light sections	TON	1,340	3,480	4,820
05410.10	**METAL FRAMING**				
0100	Furring channel, galvanized				
0110	Beams and columns, 3/4"				
0120	12" o.c.	S.F.	5.81	0.39	6.20
0140	16" o.c.	"	5.28	0.30	5.58
0150	Walls, 3/4"				
0160	12" o.c.	S.F.	2.90	0.39	3.29
0170	16" o.c.	"	2.42	0.30	2.72
0172	24" o.c.	"	1.93	0.22	2.15
0173	1-1/2"				
0174	12" o.c.	S.F.	2.90	0.66	3.56
0175	16" o.c.	"	2.42	0.49	2.91
0176	24" o.c.	"	1.93	0.34	2.27

		UNIT	LABOR	MAT.	TOTAL
05410.10	**METAL FRAMING, Cont'd...**				
0177	Stud, load bearing				
0178	16" o.c.				
0179	16 ga.				
0180	2-1/2"	S.F.	2.58	1.21	3.79
0190	3-5/8"	"	2.58	1.43	4.01
0200	4"	"	2.58	1.48	4.06
0220	6"	"	2.90	1.87	4.77
0280	18 ga.				
0300	2-1/2"	S.F.	2.58	0.99	3.57
0310	3-5/8"	"	2.58	1.21	3.79
0320	4"	"	2.58	1.26	3.84
0330	6"	"	2.90	1.59	4.49
0350	8"	"	2.90	1.92	4.82
0360	20 ga.				
0370	2-1/2"	S.F.	2.58	0.55	3.13
0390	3-5/8"	"	2.58	0.66	3.24
0400	4"	"	2.58	0.71	3.29
0420	6"	"	2.90	0.88	3.78
0440	8"	"	2.90	1.04	3.94
0480	24" o.c.				
0490	16 ga.				
0500	2-1/2"	S.F.	2.23	0.82	3.05
0510	3-5/8"	"	2.23	0.99	3.22
0520	4"	"	2.23	1.04	3.27
0530	6"	"	2.42	1.26	3.68
0540	8"	"	2.42	1.59	4.01
0545	18 ga.				
0550	2-1/2"	S.F.	2.23	0.66	2.89
0560	3-5/8"	"	2.23	0.77	3.00
0570	4"	"	2.23	0.82	3.05
0580	6"	"	2.42	1.04	3.46
0590	8"	"	2.42	1.26	3.68
0595	20 ga.				
0600	2-1/2"	S.F.	2.23	0.44	2.67
0610	3-5/8"	"	2.23	0.49	2.72
0620	4"	"	2.23	0.55	2.78
0630	6"	"	2.42	0.71	3.13
0640	8"	"	2.42	0.88	3.30
05510.10	**STAIRS**				
1000	Stock unit, steel, complete, per riser				
1010	Tread				
1020	3'-6" wide	EA.	73.00	140	213
1040	4' wide	"	83.00	160	243
1060	5' wide	"	97.00	180	277
1200	Metal pan stair, cement filled, per riser				
1220	3'-6" wide	EA.	58.00	150	208
1240	4' wide	"	65.00	170	235
1260	5' wide	"	73.00	190	263
1280	Landing, steel pan	S.F.	14.50	58.00	72.50
1300	Cast iron tread, steel stringers, stock units, per riser				
1310	Tread				

		UNIT	LABOR	MAT.	TOTAL
05510.10	**STAIRS, Cont'd...**				
1320	3'-6" wide	EA.	73.00	260	333
1340	4' wide	"	83.00	300	383
1360	5' wide	"	97.00	360	457
1400	Stair treads, abrasive, 12" x 3'-6"				
1410	Cast iron				
1420	3/8"	EA.	29.00	140	169
1440	1/2"	"	29.00	180	209
1450	Cast aluminum				
1460	5/16"	EA.	29.00	170	199
1480	3/8"	"	29.00	180	209
1500	1/2"	"	29.00	210	239
05515.10	**LADDERS**				
0100	Ladder, 18" wide				
0110	With cage	L.F.	38.75	94.00	133
0120	Without cage	"	29.00	59.00	88.00
05520.10	**RAILINGS**				
0080	Railing, pipe				
0090	1-1/4" diameter, welded steel				
0095	2-rail				
0100	Primed	L.F.	11.50	29.75	41.25
0120	Galvanized	"	11.50	38.00	49.50
0130	3-rail				
0140	Primed	L.F.	14.50	38.00	52.50
0160	Galvanized	"	14.50	49.25	63.75
0170	Wall mounted, single rail, welded steel				
0180	Primed	L.F.	8.94	19.75	28.69
0200	Galvanized	"	8.94	25.75	34.69
0210	1-1/2" diameter, welded steel				
0215	2-rail				
0220	Primed	L.F.	11.50	32.25	43.75
0240	Galvanized	"	11.50	41.75	53.25
0245	3-rail				
0250	Primed	L.F.	14.50	40.50	55.00
0260	Galvanized	"	14.50	53.00	67.50
0270	Wall mounted, single rail, welded steel				
0280	Primed	L.F.	8.94	21.50	30.44
0300	Galvanized	"	8.94	28.00	36.94
0960	2" diameter, welded steel				
0980	2-rail				
1000	Primed	L.F.	13.00	40.50	53.50
1020	Galvanized	"	13.00	53.00	66.00
1030	3-rail				
1040	Primed	L.F.	16.50	51.00	67.50
1070	Galvanized	"	16.50	66.00	82.50
1075	Wall mounted, single rail, welded steel				
1080	Primed	L.F.	9.68	23.25	32.93
1100	Galvanized	"	9.68	30.25	39.93
05580.10	**METAL SPECIALTIES**				
0060	Kick plate				
0080	4" high x 1/4" thick				
0100	Primed	L.F.	11.50	7.31	18.81

		UNIT	LABOR	MAT.	TOTAL
05580.10	**METAL SPECIALTIES, Cont'd...**				
0120	Galvanized	L.F.	11.50	8.30	19.80
0130	6" high x 1/4" thick				
0140	Primed	L.F.	13.00	7.92	20.92
0160	Galvanized	"	13.00	9.40	22.40
0200	Fire Escape (10'-12' high)				
0210	Landing with fixed stair, 3'-6" wide	EA.	1,160	5,060	6,220
0220	4'-6" wide	"	1,160	5,720	6,880
05700.10	**ORNAMENTAL METAL**				
1020	Railings, vertical square bars, 6" o.c., with shaped top rails				
1040	Steel	L.F.	29.00	88.00	117
1060	Aluminum	"	29.00	110	139
1080	Bronze	"	38.75	190	229
1100	Stainless steel	"	38.75	180	219
1200	Laminated metal or wood handrails with metal supports				
1220	2-1/2" round or oval shape	L.F.	29.00	280	309
2020	Grilles and louvers				
2040	Fixed type louvers				
2060	4 through 10 sf	S.F.	9.68	28.25	37.93
2080	Over 10 sf	"	7.26	33.25	40.51
2200	Movable type louvers				
2220	4 through 10 sf	S.F.	9.68	33.25	42.93
2240	Over 10 sf	"	7.26	36.75	44.01
3000	Aluminum louvers				
3010	Residential use, fixed type, with screen				
3020	8" x 8"	EA.	29.00	17.25	46.25
3060	12" x 12"	"	29.00	19.00	48.00
3080	12" x 18"	"	29.00	22.75	51.75
3100	14" x 24"	"	29.00	32.75	61.75
3120	18" x 24"	"	29.00	36.75	65.75
3140	30" x 24"	"	32.25	50.00	82.25

		UNIT	COST
05999.10	**STRUCTURAL STEEL FRAME**		
1000	To 30 Ton 20' Span	S.F.	11.21
1010	24' Span	"	11.82
1020	28' Span	"	11.94
1030	32' Span	"	14.32
1040	36' Span	"	15.96
2010	44' Span	"	18.20
2020	48' Span	"	19.55
2030	52' Span	"	21.82
2050	60' Span	"	25.37
3000	Deduct for Over 30 Ton	PCT.	5.00
05999.40	**OPEN WEB JOISTS**		
1000	To 20 Ton, 20' Span	S.F.	4.15
1010	24' Span	"	4.18
1030	32' Span	"	4.27
2000	40' Span	"	4.91
2010	48' Span	"	5.80
2030	56' Span	"	6.76
2040	60' Span	"	7.26
3000	Deduct for Over 20 Ton	PCT.	10.00
05999.50	**METAL DECKING**		
1000	1/2" Deep - Ribbed - Baked Enamel - 18 Ga	S.F.	4.42
1020	22 Ga	"	3.71
2000	3" Deep - Ribbed - Baked Enamel - 18 Ga	"	8.79
2020	22 Ga	"	7.74
3000	4 1/2" Deep - Ribbed - Baked Enamel - 16 Ga	"	11.58
3010	18 Ga	"	10.00
3020	20 Ga	"	9.31
4000	3" Deep - Cellular - 18 Ga	"	11.29
4010	16 Ga	"	13.93
4020	4 1/2" Deep - Cellular - 18 Ga	"	15.96
4030	16 Ga	"	18.29
4040	Add for Galvanized	PCT	16.50
4050	Corrugated Black Standard .015	S.F.	2.26
4060	Heavy Duty - 26 Ga	"	2.44
4070	S. Duty - 24 Ga	"	3.20
4080	22 Ga	"	3.32
5000	Add for Galvanized	PCT.	15.00
05999.60	**METAL SIDING AND ROOFING**		
1000	Aluminum- Anodized	S.F.	5.61
1010	Porcelainized	"	11.16
1020	Corrugated	"	3.80
1030	Enamel, Baked Ribbed	"	5.94
1040	24 Ga Ribbed	"	4.07
1050	Acrylic	"	6.96
2000	Porcelain Ribbed	"	7.67
2010	Galvanized- Corrugated	"	3.83
2020	Plastic Faced	"	8.34
2030	Protected Metal	"	10.16
2040	Add for Liner Panels	"	3.15
2050	Add for Insulation	"	0.50

TABLE OF CONTENTS   PAGE

	UNIT	LABOR	MAT.	TOTAL
06110.10 BLOCKING				
1215 Wood construction				
1220 Walls				
1230 2x4	L.F.	2.90	0.41	3.31
1240 2x6	"	3.26	0.61	3.87
1250 2x8	"	3.48	0.84	4.32
1260 2x10	"	3.72	1.02	4.74
1270 2x12	"	4.01	1.87	5.88
1280 Ceilings				
1290 2x4	L.F.	3.26	0.41	3.67
1300 2x6	"	3.72	0.61	4.33
1310 2x8	"	4.01	0.84	4.85
1320 2x10	"	4.35	1.02	5.37
1330 2x12	"	4.74	1.87	6.61
06110.20 CEILING FRAMING				
1000 Ceiling joists				
1070 16" o.c.				
1080 2x4	S.F.	1.00	0.49	1.49
1090 2x6	"	1.04	0.74	1.78
1100 2x8	"	1.08	1.04	2.12
1110 2x10	"	1.13	1.21	2.34
1120 2x12	"	1.18	2.20	3.38
1130 24" o.c.				
1140 2x4	S.F.	0.82	0.38	1.20
1150 2x6	"	0.87	0.60	1.47
1160 2x8	"	0.91	0.88	1.79
1170 2x10	"	0.96	1.04	2.00
1180 2x12	"	1.02	2.64	3.66
1200 Headers and nailers				
1210 2x4	L.F.	1.68	0.41	2.09
1220 2x6	"	1.74	0.61	2.35
1230 2x8	"	1.86	0.84	2.70
1240 2x10	"	2.00	1.04	3.04
1250 2x12	"	2.17	1.26	3.43
1300 Sister joists for ceilings				
1310 2x4	L.F.	3.72	0.41	4.13
1320 2x6	"	4.35	0.61	4.96
1330 2x8	"	5.22	0.84	6.06
1340 2x10	"	6.52	1.04	7.56
1350 2x12	"	8.70	1.26	9.96
06110.30 FLOOR FRAMING				
1000 Floor joists				
1180 16" o.c.				
1190 2x6	S.F.	0.87	0.64	1.51
1200 2x8	"	0.88	0.90	1.78
1220 2x10	"	0.90	1.10	2.00
1230 2x12	"	0.93	1.37	2.30
1240 2x14	"	0.96	3.12	4.08
1250 3x6	"	0.90	2.12	3.02
1260 3x8	"	0.93	2.76	3.69
1270 3x10	"	0.96	3.44	4.40
1280 3x12	"	1.00	4.11	5.11

		UNIT	LABOR	MAT.	TOTAL
06110.30	**FLOOR FRAMING, Cont'd...**				
1290	3x14	S.F.	1.04	4.84	5.88
1300	4x6	"	0.90	2.75	3.65
1310	4x8	"	0.93	3.68	4.61
1320	4x10	"	0.96	4.58	5.54
1330	4x12	"	1.00	5.42	6.42
1340	4x14	"	1.04	6.43	7.47
2000	Sister joists for floors				
2010	2x4	L.F.	3.26	0.41	3.67
2020	2x6	"	3.72	0.61	4.33
2030	2x8	"	4.35	0.84	5.19
2040	2x10	"	5.22	1.02	6.24
2050	2x12	"	6.52	1.87	8.39
2060	3x6	"	5.22	2.12	7.34
2070	3x8	"	5.80	2.76	8.56
2080	3x10	"	6.52	3.44	9.96
2090	3x12	"	7.45	4.11	11.56
2100	4x6	"	5.22	2.67	7.89
2110	4x8	"	5.80	3.68	9.48
2120	4x10	"	6.52	4.58	11.10
2130	4x12	"	7.45	5.42	12.87
06110.40	**FURRING**				
1100	Furring, wood strips				
1102	Walls				
1105	On masonry or concrete walls				
1107	1x2 furring				
1110	12" o.c.	S.F.	1.63	0.33	1.96
1120	16" o.c.	"	1.49	0.28	1.77
1130	24" o.c.	"	1.37	0.26	1.63
1135	1x3 furring				
1140	12" o.c.	S.F.	1.63	0.42	2.05
1150	16" o.c.	"	1.49	0.37	1.86
1160	24" o.c.	"	1.37	0.29	1.66
1165	On wood walls				
1167	1x2 furring				
1170	12" o.c.	S.F.	1.16	0.33	1.49
1180	16" o.c.	"	1.04	0.28	1.32
1190	24" o.c.	"	0.94	0.26	1.20
1195	1x3 furring				
1200	12" o.c.	S.F.	1.16	0.42	1.58
1210	16" o.c.	"	1.04	0.37	1.41
1220	24" o.c.	"	0.94	0.29	1.23
06110.50	**ROOF FRAMING**				
1000	Roof framing				
1005	Rafters, gable end				
1008	0-2 pitch (flat to 2-in-12)				
1070	16" o.c.				
1080	2x6	S.F.	0.93	0.74	1.67
1090	2x8	"	0.96	1.03	1.99
1100	2x10	"	1.00	1.26	2.26
1110	2x12	"	1.04	2.20	3.24
1120	24" o.c.				

		UNIT	LABOR	MAT.	TOTAL
1120	24" o.c.				
1130	2x6	S.F.	0.79	0.42	1.21
1140	2x8	"	0.81	0.85	1.66
1150	2x10	"	0.84	1.00	1.84
1160	2x12	"	0.87	1.81	2.68
1165	4-6 pitch (4-in-12 to 6-in-12)				
1220	16" o.c.				
1230	2x6	S.F.	0.96	0.74	1.70
1240	2x8	"	1.00	1.21	2.21
1250	2x10	"	1.04	1.32	2.36
1260	2x12	"	1.08	1.96	3.04
1270	24" o.c.				
1280	2x6	S.F.	0.81	0.61	1.42
1290	2x8	"	0.84	0.99	1.83
1300	2x10	"	0.90	1.08	1.98
1310	2x12	"	1.00	1.65	2.65
1315	8-12 pitch (8-in-12 to 12-in-12)				
1380	16" o.c.				
1390	2x6	S.F.	1.00	0.82	1.82
1400	2x8	"	1.04	1.32	2.36
1410	2x10	"	1.08	1.48	2.56
1420	2x12	"	1.13	2.14	3.27
1430	24" o.c.				
1440	2x6	S.F.	0.84	0.63	1.47
1450	2x8	"	0.87	1.04	1.91
1460	2x10	"	0.90	1.21	2.11
1470	2x12	"	0.93	1.92	2.85
2000	Ridge boards				
2010	2x6	L.F.	2.61	0.61	3.22
2020	2x8	"	2.90	0.82	3.72
2030	2x10	"	3.26	1.04	4.30
2040	2x12	"	3.72	1.92	5.64
3000	Hip rafters				
3010	2x6	L.F.	1.86	0.61	2.47
3020	2x8	"	1.93	0.82	2.75
3030	2x10	"	2.00	1.04	3.04
3040	2x12	"	2.08	1.92	4.00
3180	Jack rafters				
3190	4-6 pitch (4-in-12 to 6-in-12)				
3200	16" o.c.				
3210	2x6	S.F.	1.53	0.75	2.28
3220	2x8	"	1.58	1.26	2.84
3230	2x10	"	1.68	1.39	3.07
3240	2x12	"	1.74	2.02	3.76
3250	24" o.c.				
3260	2x6	S.F.	1.18	0.61	1.79
3270	2x8	"	1.21	1.04	2.25
3280	2x10	"	1.27	1.26	2.53
3290	2x12	"	1.30	1.70	3.00
3295	8-12 pitch (8-in-12 to 12-in-12)				
3300	16" o.c.				
3310	2x6	S.F.	1.63	1.21	2.84

		UNIT	LABOR	MAT.	TOTAL
06110.50	**ROOF FRAMING, Cont'd...**				
3320	2x8	S.F.	1.68	1.47	3.15
3330	2x10	"	1.74	2.20	3.94
3340	2x12	"	1.80	3.02	4.82
3350	24" o.c.				
3360	2x6	S.F.	1.24	0.93	2.17
3370	2x8	"	1.27	1.26	2.53
3380	2x10	"	1.30	1.92	3.22
3390	2x12	"	1.33	2.69	4.02
4980	Sister rafters				
5000	2x4	L.F.	3.72	0.39	4.11
5010	2x6	"	4.35	0.61	4.96
5020	2x8	"	5.22	0.84	6.06
5030	2x10	"	6.52	1.04	7.56
5040	2x12	"	8.70	1.87	10.57
5050	Fascia boards				
5060	2x4	L.F.	2.61	0.39	3.00
5070	2x6	"	2.61	0.61	3.22
5080	2x8	"	2.90	0.84	3.74
5090	2x10	"	2.90	1.02	3.92
5100	2x12	"	3.26	1.87	5.13
7980	Cant strips				
7985	Fiber				
8000	3x3	L.F.	1.49	0.33	1.82
8020	4x4	"	1.58	0.42	2.00
8030	Wood				
8040	3x3	L.F.	1.58	1.68	3.26
06110.60	**SLEEPERS**				
0960	Sleepers, over concrete				
1090	16" o.c.				
1100	1x2	S.F.	1.04	0.21	1.25
1120	1x3	"	1.04	0.30	1.34
1140	2x4	"	1.24	0.62	1.86
1160	2x6	"	1.30	0.94	2.24
06110.65	**SOFFITS**				
0980	Soffit framing				
1000	2x3	L.F.	3.72	0.33	4.05
1020	2x4	"	4.01	0.41	4.42
1030	2x6	"	4.35	0.61	4.96
1040	2x8	"	4.74	0.84	5.58
06110.70	**WALL FRAMING**				
0960	Framing wall, studs				
1110	16" o.c.				
1120	2x3	S.F.	0.81	0.35	1.16
1140	2x4	"	0.81	0.49	1.30
1150	2x6	"	0.87	0.71	1.58
1160	2x8	"	0.90	1.10	2.00
1165	24" o.c.				
1170	2x3	S.F.	0.70	0.27	0.97
1180	2x4	"	0.70	0.38	1.08
1190	2x6	"	0.74	0.60	1.34
1200	2x8	"	0.76	0.82	1.58

06110.70	WALL FRAMING, Cont'd...	UNIT	LABOR	MAT.	TOTAL
1480	Plates, top or bottom				
1500	2x3	L.F.	1.53	0.33	1.86
1510	2x4	"	1.63	0.41	2.04
1520	2x6	"	1.74	0.61	2.35
1530	2x8	"	1.86	0.84	2.70
2000	Headers, door or window				
2044	2x8				
2046	Single				
2050	4' long	EA.	32.50	3.38	35.88
2060	8' long	"	40.25	6.76	47.01
2065	Double				
2070	4' long	EA.	37.25	6.76	44.01
2080	8' long	"	47.50	13.50	61.00
2134	2x12				
2138	Single				
2140	6' long	EA.	40.25	7.42	47.67
2150	12' long	"	52.00	14.50	66.50
2155	Double				
2160	6' long	EA.	47.50	14.50	62.00
2170	12' long	"	58.00	29.00	87.00
06115.10	FLOOR SHEATHING				
1980	Sub-flooring, plywood, CDX				
2000	1/2" thick	S.F.	0.65	0.69	1.34
2020	5/8" thick	"	0.74	1.10	1.84
2080	3/4" thick	"	0.87	1.26	2.13
2090	Structural plywood				
2100	1/2" thick	S.F.	0.65	0.77	1.42
2120	5/8" thick	"	0.74	1.25	1.99
2140	3/4" thick	"	0.80	1.37	2.17
5990	Underlayment				
6000	Hardboard, 1/4" tempered	S.F.	0.65	0.61	1.26
6010	Plywood, CDX				
6020	3/8" thick	S.F.	0.65	0.77	1.42
6040	1/2" thick	"	0.69	0.88	1.57
6060	5/8" thick	"	0.74	1.04	1.78
6080	3/4" thick	"	0.80	1.26	2.06
06115.20	ROOF SHEATHING				
0080	Sheathing				
0090	Plywood, CDX				
1000	3/8" thick	S.F.	0.67	0.77	1.44
1020	1/2" thick	"	0.69	0.88	1.57
1040	5/8" thick	"	0.74	1.04	1.78
1060	3/4" thick	"	0.80	1.26	2.06
1080	Structural plywood				
2040	3/8" thick	S.F.	0.67	0.55	1.22
2060	1/2" thick	"	0.69	0.71	1.40
2080	5/8" thick	"	0.74	0.88	1.62
2100	3/4" thick	"	0.80	1.02	1.82
06115.30	WALL SHEATHING				
0980	Sheathing				
0990	Plywood, CDX				

		UNIT	LABOR	MAT.	TOTAL
06115.30	**WALL SHEATHING, Cont'd...**				
1000	3/8" thick	S.F.	0.77	0.77	1.54
1020	1/2" thick	"	0.80	0.88	1.68
1040	5/8" thick	"	0.87	1.04	1.91
1060	3/4" thick	"	0.94	1.26	2.20
3000	Waferboard				
3020	3/8" thick	S.F.	0.77	0.71	1.48
3040	1/2" thick	"	0.80	0.95	1.75
3060	5/8" thick	"	0.87	1.17	2.04
3080	3/4" thick	"	0.94	1.21	2.15
4100	Structural plywood				
4120	3/8" thick	S.F.	0.77	0.55	1.32
4140	1/2" thick	"	0.80	0.77	1.57
4160	5/8" thick	"	0.87	1.25	2.12
4180	3/4" thick	"	0.94	1.37	2.31
7000	Gypsum, 1/2" thick	"	0.80	0.35	1.15
8000	Asphalt impregnated fiberboard, 1/2" thick	"	0.80	0.72	1.52
06125.10	**WOOD DECKING**				
0090	Decking, T&G solid				
0095	Cedar				
0100	3" thick	S.F.	1.30	9.20	10.50
0120	4" thick	"	1.39	11.25	12.64
1030	Fir				
1040	3" thick	S.F.	1.30	3.85	5.15
1060	4" thick	"	1.39	4.69	6.08
1080	Southern yellow pine				
2000	3" thick	S.F.	1.49	3.79	5.28
2020	4" thick	"	1.60	4.01	5.61
3120	White pine				
3140	3" thick	S.F.	1.30	4.78	6.08
3160	4" thick	"	1.39	6.31	7.70
06130.10	**HEAVY TIMBER**				
1000	Mill framed structures				
1010	Beams to 20' long				
1020	Douglas fir				
1040	6x8	L.F.	6.72	6.12	12.84
1042	6x10	"	6.95	7.22	14.17
1044	6x12	"	7.47	8.64	16.11
1046	6x14	"	7.76	10.25	18.01
1048	6x16	"	8.07	11.25	19.32
1060	8x10	"	6.95	9.58	16.53
1070	8x12	"	7.47	11.25	18.72
1080	8x14	"	7.76	13.00	20.76
1090	8x16	"	8.07	14.75	22.82
1380	Columns to 12' high				
1400	Douglas fir				
1420	6x6	L.F.	10.00	4.40	14.40
1440	8x8	"	10.00	7.54	17.54
1460	10x10	"	11.25	13.25	24.50
1480	12x12	"	11.25	16.25	27.50
0080	Posts, treated				
0100	4x4	L.F.	2.08	1.51	3.59

		UNIT	LABOR	MAT.	TOTAL
06132.10	**HEAVY TIMBER, Cont'd...**				
0120	6x6	L.F.	2.61	4.40	7.01
06190.20	**WOOD TRUSSES**				
0960	Truss, fink, 2x4 members				
0980	3-in-12 slope				
1030	5-in-12 slope				
1040	24' span	EA.	59.00	93.00	152
1050	28' span	"	61.00	100	161
1055	30' span	"	63.00	110	173
1060	32' span	"	63.00	110	173
1070	40' span	"	67.00	150	217
1074	Gable, 2x4 members				
1078	5-in-12 slope				
1080	24' span	EA.	59.00	110	169
1100	28' span	"	61.00	130	191
1120	30' span	"	63.00	140	203
1160	36' span	"	65.00	150	215
1180	40' span	"	67.00	160	227
1190	King post type, 2x4 members				
2000	4-in-12 slope				
2040	16' span	EA.	55.00	65.00	120
2060	18' span	"	56.00	70.00	126
2080	24' span	"	59.00	74.00	133
2120	30' span	"	63.00	100	163
2160	38' span	"	65.00	130	195
2180	42' span	"	70.00	160	230
06190.30	**LAMINATED BEAMS**				
0010	Parallel strand beams 3-1/2" wide x				
0020	9-1/2"	L.F.	2.88	8.41	11.29
0030	11-1/4"	"	2.98	9.07	12.05
0040	11-7/8"	"	3.10	10.00	13.10
0050	14"	"	3.66	12.25	15.91
0060	16"	"	4.03	14.75	18.78
0070	18"	"	4.48	17.25	21.73
1000	Laminated veneer beams, 1-3/4" wide x				
1010	11-7/8"	L.F.	3.10	5.88	8.98
1020	14"	"	3.66	7.31	10.97
1030	16"	"	4.03	7.20	11.23
1040	18"	"	4.48	9.18	13.66
2000	Laminated strand beams, 1-3/4" wide x				
2010	9-1/2"	L.F.	2.88	3.79	6.67
2020	11-7/8"	"	3.10	4.34	7.44
2030	14"	"	3.66	5.06	8.72
2040	16"	"	4.03	5.77	9.80
2050	3-1/2" wide x				
2060	9-1/2"	L.F.	2.88	6.60	9.48
2070	11-7/8"	"	3.10	8.41	11.51
2080	14"	"	3.66	10.25	13.91
2090	16"	"	4.03	12.00	16.03
3000	Gluelam beam, 3-1/2" wide x				
3010	10"	L.F.	2.88	10.00	12.88
3020	12"	"	3.36	11.75	15.11

		UNIT	LABOR	MAT.	TOTAL
06190.30	**LAMINATED BEAMS, Cont'd...**				
3030	15"	L.F.	3.84	14.00	17.84
3040	5-1/2" wide x				
3050	10"	L.F.	2.88	16.00	18.88
3060	16"	"	4.03	25.25	29.28
3070	20"	"	4.74	29.75	34.49
06200.10	**FINISH CARPENTRY**				
0070	Mouldings and trim				
0980	Apron, flat				
1000	9/16 x 2	L.F.	2.61	1.37	3.98
1010	9/16 x 3-1/2	"	2.74	2.69	5.43
1015	Base				
1020	Colonial				
1022	7/16 x 2-1/4	L.F.	2.61	1.65	4.26
1024	7/16 x 3	"	2.61	2.03	4.64
1026	7/16 x 3-1/4	"	2.61	2.20	4.81
1028	9/16 x 3	"	2.74	2.14	4.88
1030	9/16 x 3-1/4	"	2.74	2.25	4.99
1034	11/16 x 2-1/4	"	2.90	2.36	5.26
1035	Ranch				
1036	7/16 x 2-1/4	L.F.	2.61	1.81	4.42
1038	7/16 x 3-1/4	"	2.61	2.20	4.81
1039	9/16 x 2-1/4	"	2.74	1.98	4.72
1041	9/16 x 3	"	2.74	2.03	4.77
1043	9/16 x 3-1/4	"	2.74	2.20	4.94
1050	Casing				
1060	11/16 x 2-1/2	L.F.	2.37	1.69	4.06
1070	11/16 x 3-1/2	"	2.48	1.92	4.40
1180	Chair rail				
1200	9/16 x 2-1/2	L.F.	2.61	1.80	4.41
1210	9/16 x 3-1/2	"	2.61	2.53	5.14
1250	Closet pole				
1300	1-1/8" dia.	L.F.	3.48	1.21	4.69
1310	1-5/8" dia.	"	3.48	1.73	5.21
1340	Cove				
1500	9/16 x 1-3/4	L.F.	2.61	1.37	3.98
1510	11/16 x 2-3/4	"	2.61	1.92	4.53
1550	Crown				
1600	9/16 x 1-5/8	L.F.	3.48	1.81	5.29
1620	11/16 x 3-5/8	"	4.35	2.14	6.49
1640	11/16 x 5-1/4	"	5.22	3.57	8.79
1680	Drip cap				
1700	1-1/16 x 1-5/8	L.F.	2.61	1.92	4.53
1780	Glass bead				
1800	3/8 x 3/8	L.F.	3.26	0.66	3.92
1840	5/8 x 5/8	"	3.26	0.84	4.10
1860	3/4 x 3/4	"	3.26	0.93	4.19
1880	Half round				
1900	1/2	L.F.	2.08	0.77	2.85
1910	5/8	"	2.08	1.04	3.12
1920	3/4	"	2.08	1.32	3.40
1980	Lattice				

06200.10	FINISH CARPENTRY, Cont'd...	UNIT	LABOR	MAT.	TOTAL
2000	1/4 x 7/8	L.F.	2.08	0.60	2.68
2010	1/4 x 1-1/8	"	2.08	0.66	2.74
2030	1/4 x 1-3/4	"	2.08	0.80	2.88
2040	1/4 x 2	"	2.08	0.93	3.01
2080	Ogee molding				
2100	5/8 x 3/4	L.F.	2.61	1.26	3.87
2110	11/16 x 1-1/8	"	2.61	1.92	4.53
2120	11/16 x 1-3/8	"	2.61	2.36	4.97
2180	Parting bead				
2200	3/8 x 7/8	L.F.	3.26	1.04	4.30
2300	Quarter round				
2301	1/4 x 1/4	L.F.	2.08	0.33	2.41
2303	3/8 x 3/8	"	2.08	0.49	2.57
2305	1/2 x 1/2	"	2.08	0.68	2.76
2307	11/16 x 11/16	"	2.26	0.68	2.94
2309	3/4 x 3/4	"	2.26	1.26	3.52
2311	1-1/16 x 1-1/16	"	2.37	0.93	3.30
2380	Railings, balusters				
2400	1-1/8 x 1-1/8	L.F.	5.22	3.46	8.68
2410	1-1/2 x 1-1/2	"	4.74	4.01	8.75
2480	Screen moldings				
2500	1/4 x 3/4	L.F.	4.35	0.82	5.17
2510	5/8 x 5/16	"	4.35	1.04	5.39
2580	Shoe				
2600	7/16 x 11/16	L.F.	2.08	1.04	3.12
2605	Sash beads				
2620	1/2 x 7/8	L.F.	4.35	1.37	5.72
2640	5/8 x 7/8	"	4.74	1.48	6.22
2760	Stop				
2780	5/8 x 1-5/8				
2800	Colonial	L.F.	3.26	0.71	3.97
2810	Ranch	"	3.26	0.71	3.97
2880	Stools				
2900	11/16 x 2-1/4	L.F.	5.80	3.30	9.10
2910	11/16 x 2-1/2	"	5.80	3.41	9.21
2920	11/16 x 5-1/4	"	6.52	3.57	10.09
4000	Exterior trim, casing, select pine, 1x3	"	2.61	2.36	4.97
4010	Douglas fir				
4020	1x3	L.F.	2.61	1.10	3.71
4040	1x4	"	2.61	1.37	3.98
4060	1x6	"	2.90	1.81	4.71
4100	1x8	"	3.26	2.47	5.73
5000	Cornices, white pine, #2 or better				
5040	1x4	L.F.	2.61	0.82	3.43
5080	1x8	"	3.07	1.65	4.72
5120	1x12	"	3.48	2.75	6.23
8600	Shelving, pine				
8620	1x8	L.F.	4.01	1.18	5.19
8640	1x10	"	4.17	1.60	5.77
8660	1x12	"	4.35	2.03	6.38
8800	Plywood shelf, 3/4", with edge band, 12" wide	"	5.22	2.20	7.42
8840	Adjustable shelf, and rod, 12" wide				

		UNIT	LABOR	MAT.	TOTAL
06200.10	**FINISH CARPENTRY, Cont'd...**				
8860	3' to 4' long	EA.	13.00	17.50	30.50
8880	5' to 8' long	"	17.50	33.00	50.50
8900	Prefinished wood shelves with brackets and supports				
8905	8" wide				
8910	3' long	EA.	13.00	52.00	65.00
8922	4' long	"	13.00	59.00	72.00
8924	6' long	"	13.00	86.00	99.00
8930	10" wide				
8940	3' long	EA.	13.00	57.00	70.00
8942	4' long	"	13.00	83.00	96.00
8946	6' long	"	13.00	94.00	107
06220.10	**MILLWORK**				
0070	Countertop, laminated plastic				
0080	25" x 7/8" thick				
0099	Minimum	L.F.	13.00	13.75	26.75
0100	Average	"	17.50	26.00	43.50
0110	Maximum	"	20.75	38.50	59.25
0115	25" x 1-1/4" thick				
0120	Minimum	L.F.	17.50	16.50	34.00
0130	Average	"	20.75	33.25	54.00
0140	Maximum	"	26.00	50.00	76.00
0160	Add for cutouts	EA.	32.50		32.50
0165	Backsplash, 4" high, 7/8" thick	L.F.	10.50	18.25	28.75
2000	Plywood, sanded, A-C				
2020	1/4" thick	S.F.	1.74	1.18	2.92
2040	3/8" thick	"	1.86	1.29	3.15
2060	1/2" thick	"	2.00	1.43	3.43
2070	A-D				
2080	1/4" thick	S.F.	1.74	1.10	2.84
2090	3/8" thick	"	1.86	1.24	3.10
2100	1/2" thick	"	2.00	1.37	3.37
2500	Base cabinets, 34-1/2" high, 24" deep, hardwood, no tops				
2540	Minimum	L.F.	20.75	180	201
2560	Average	"	26.00	210	236
2580	Maximum	"	34.75	230	265
2600	Wall cabinets				
2640	Minimum	L.F.	17.50	55.00	72.50
2660	Average	"	20.75	74.00	94.75
2680	Maximum	"	26.00	94.00	120
06300.10	**WOOD TREATMENT**				
1000	Creosote preservative treatment				
1020	8 lb/cf	B.F.			0.59
1040	10 lb/cf	"			0.71
1060	Salt preservative treatment				
1070	Oil borne				
1080	Minimum	B.F.			0.44
1100	Maximum	"			0.71
1120	Water borne				
1140	Minimum	B.F.			0.33
1150	Maximum	"			0.55
1200	Fire retardant treatment				

		UNIT	LABOR	MAT.	TOTAL
06300.10	**WOOD TREATMENT, Cont'd...**				
1220	Minimum	B.F.			0.73
1240	Maximum	"			0.88
1300	Kiln dried, softwood, add to framing costs				
1320	1" thick	B.F.			0.21
1360	3" thick	"			0.44
06430.10	**STAIRWORK**				
0080	Risers, 1x8, 42" wide				
0100	White oak	EA.	26.00	38.50	64.50
0120	Pine	"	26.00	33.00	59.00
0130	Treads, 1-1/16" x 9-1/2" x 42"				
0140	White oak	EA.	32.50	46.25	78.75
06440.10	**COLUMNS**				
0980	Column, hollow, round wood				
0990	12" diameter				
1000	10' high	EA.	74.00	680	754
1080	16' high	"	110	1,240	1,350
2000	24" diameter				
2020	16' high	EA.	110	2,830	2,940
2060	20' high	"	120	3,960	4,080
2100	24' high	"	120	4,550	4,670

		UNIT	COST
06999.10	**ROUGH CARPENTRY**		
1000	LIGHT FRAMING AND SHEATHING		
1100	Joists and Headers - Floor Area - 16" O.C.		
1110	2" x 6" Joists - with Headers & Bridging	S.F.	2.88
1120	2" x 8" Joists	"	3.54
1130	2" x 10" Joists	"	3.96
1140	2" x 12" Joists	"	4.72
1150	Add for Ceiling Joists, 2nd Floor and Above	"	0.09
1160	Add for Sloped Installation	"	0.16
1200	Studs, Plates and Framing - 8' Wall Height - 16" O.C.	"	1.69
1210	2" x 3" Stud Wall - Non-Bearing (Single Top Plate)		
1220	2" x 4" Stud Wall - Bearing (Double Top Plate)	S.F.	2.05
1230	2" x 4" Stud Wall - Non-Bearing (Single Top Plate)	"	1.48
1240	2" x 6" Stud Wall - Bearing (Double Top Plate)	"	2.80
1250	2" x 6" Stud Wall - Non-Bearing (Single Top Plate)	"	2.02
1300	Add for Stud Wall - 12" O.C.	"	0.09
1310	Deduct for Stud Wall - 24" O.C.	"	0.25
1320	Add for Bolted Plates or Sills	"	0.13
1330	Add for Each Foot Above 8'	"	0.07
1340	Add to Above for Fire Stops, Fillers and Nailers	"	0.54
1350	Add for Soffits and Suspended Framing	"	0.75
1360	Bridging - 1" x 3" Wood Diagonal	EA.	2.92
1370	2" x 8" Solid	"	3.22
1400	Rafters		
1410	2" x 4" Rafter (Incl Bracing) 3 - 12 Slope	S.F.	2.13
1420	4 - 12 Slope	"	2.13
1430	5 - 12 Slope	"	2.41
1440	6 - 12 Slope	"	2.80
1500	Add for Hip-and-Valley Type	"	0.51
1510	Add for 1' 0" Overhang - Total Area of Roof	"	0.19
1520	2" x 6" Rafter (Incl Bracing) 3-12 Slope	"	2.67
1530	4-12 Slope	"	2.70
1540	5-12 Slope	"	2.80
1550	6-12 Slope	"	2.92
1560	Add for Hip - and - Valley Type	"	0.61
1600	Stairs	EA.	540
1700	Sub Floor Sheathing (Structural)		
1710	1" x 8" and 1" x 10" #3 Pine	S.F.	1.83
1720	1/2" x 4' x 8' CD Plywood - Exterior	"	1.48
1730	5/8" x 4' x 8'	"	1.80
1740	3/4" x 4' x 8'	"	2.45
1800	Floor Sheathing (Over Sub Floor)		
1810	3/8" x 4' x 8' CD Plywood	S.F.	1.37
1820	1/2" x 4' x 8'	"	1.66
1830	5/8" x 4' x 8'	"	2.21
1840	1/2" x 4' x 8' Particle Board	"	1.73
1850	5/8" x 4' x 8'	"	2.02
1860	3/4" x 4' x 8'	"	2.34
1900	Wall Sheathing		
1910	1' x 8" and 1" x 10" - #3 Pine	S.F.	1.92
1920	3/8" x 4' x 8' CD Plywood - Exterior	"	1.41
1930	1/2" x 4' x 8'	"	1.69
1940	5/8" x 4' x 8'	"	2.28

		UNIT	COST
06999.10	**ROUGH CARPENTRY, Cont'd...**		
1950	3/4" x 4' x 8'	S.F.	2.57
1960	25/32" x 4' x 8' Fiber Board - Impregnated	"	1.58
1970	1" x 2' x 8' T&G Styrofoam	"	1.40
1980	2" x 2' x 8'	"	2.05
2000	Roof Sheathing - Flat Construction		
2010	1" x 6" and 1" x 8" - #3 Pine	S.F.	2.40
2020	1/2" x 4' x 8' CD Plywood - Exterior	"	2.22
2030	5/8" x 4' x 8'	"	1.92
2040	3/4" x 4' x 8'	"	2.58
2050	Add for Sloped Roof Construction (to 5-12 slope)	"	0.20
2060	Add for Steep Sloped Construction (over 5-12 slope)	"	0.35
2070	Add to Above Sheathing		
2080	AC or AD Plywood	S.F.	0.35
2090	10' Length Plywood	"	0.28
2100	HEAVY FRAMING		
2200	Columns and Beams - 16' Span Average - Floor Area	S.F.	8.26
2210	20' Span	"	8.63
2220	24' Span	"	10.02
2300	Deck 2" x 6" T&G - Fir Random Construction Grade	"	4.88
2310	3" x 6" T&G - Fir Random Construction Grade	"	6.51
2320	4" x 6"	"	8.26
2330	2" x 6" T&G - Red Cedar	"	5.27
2340	Add for D Grade Cedar	"	2.06
2350	2" x 6" T&G - Panelized Fir	"	4.46
3000	MISCELLANEOUS CARPENTRY		
3100	Blocking and Bucks (2" x 4" and 2" x 6")		
3110	Doors & Windows - Nailed to Concrete or Masonry	EA.	57.25
3120	Bolted to Concrete or Masonry	"	63.00
3130	Doors & Windows - Nailed to Concrete or Masonry	B.F.	2.75
3140	Bolted to Concrete or Masonry	"	3.84
3150	Roof Edges - Nailed to Wood	"	2.19
3160	Bolted to Concrete (Incl. bolts)	"	3.29
3200	Grounds and Furring		
3210	1" x 6" Fastened to Wood	L.F.	1.73
3220	1" x 4"	"	1.29
3230	1" x 3"	"	1.15
3240	1" x 3" Fastened to Concrete/Masonry - Nailed	"	1.82
3250	Gun Driven	"	1.66
3260	PreClipped	"	2.03
3270	2" x 2" Suspended Framing	"	1.82
3280	Not Suspended Framing	"	1.53
3300	Grounds and Furring		
3310	1" x 6" Fastened to Wood	B.F.	3.46
3320	1" x 4"	"	4.00
3330	1" x 3"	"	4.63
3340	1" x 3" Fastened to Concrete/Masonry - Nailed	"	6.95
3400	Gun Driven	"	6.62
3410	PreClipped	"	8.12
3420	2" x 2" Suspended Framing	"	5.42
3430	Not Suspended Framing	"	4.67
3440	Cant Strips		
3450	4" x 4" Treated and Nailed	S.F.	2.04

		UNIT	COST
06999.10	**ROUGH CARPENTRY, Cont'd...**		
3460	Treated and Bolted (Including Bolts)	S.F.	3.01
3470	6" x 6" Treated and Nailed	"	4.17
3480	Treated and Bolted (Including Bolts)	"	5.29
3600	Building Papers and Sealers		
3610	15" Felt	S.F.	0.26
3620	Polyethylene - 4 mil	"	0.20
3630	6 mil	"	0.23
3640	Sill Sealer	"	0.98
06999.20	**FINISH CARPENTRY**		
1000	FINISH SIDINGS AND FACING MATERIALS (Exterior) TO %		
1100	Boards and Beveled Sidings		
1200	Cedar Beveled - Clear Heart 1/2" x 4"	S.F.	6.53
1210	1/2" x 6"	"	5.74
1220	1/2" x 8"	"	4.90
1300	Rough Sawn 7/8" x 8"	"	5.00
1310	7/8" x 10"	"	5.17
1320	7/8" x 12"	"	5.17
1330	3/4" x 12"	"	4.48
1400	Redwood Beveled - Clear Heart		
1410	1/2" x 6"	S.F.	6.28
1430	5/8" x 10"	"	6.40
1440	3/4" x 6"	"	6.49
1450	3/4" x 8"	"	7.24
1500	3/4" x 6" Board - Clear	"	3.76
1510	3/4" Tongue & Groove	"	5.91
1520	1" Rustic Beveled - 6" and 8"	"	6.54
1530	5/4" Rustic Beveled - 6" and 8"	"	6.46
1540	Add for Metal Corners	EA.	1.79
1550	Add for Mitering Corners	"	3.82
1600	Plywood Fir AC Smooth -One Side 1/4"	S.F.	2.27
1610	3/8"	"	2.43
1620	1/2"	"	2.47
1630	5/8"	"	2.60
1640	3/4"	"	2.91
1650	5/8" - Grooved and Rough Faced	"	2.99
1700	Cedar - Rough Sawn 3/8"	"	2.89
1710	5/8"	"	2.99
1720	3/4"	"	3.56
1730	5/8" - Grooved and Rough Faced	"	3.56
1740	Add for Wood Batten Strips, 1" x 2" - 4' O.C.	"	0.45
1750	Add for Splines	"	0.45
1800	Hardboard - Paneling and Lap Siding (Primed)		
1810	3/8" Rough Textured Paneling - 4' x 8'	S.F.	2.39
1820	7/16" Grooved - 4' x 8'	"	2.29
1830	7/16" Stucco Board Paneling	"	2.51
1840	7/16" x 8" Lap Siding	"	2.87
1860	Add for Pre-Finishing	"	0.22
1900	Shingles		
1910	Wood - 16" Red Cedar - 12" to Weather - #1	S.F.	4.65
1920	#2	"	3.75
1930	#3	"	3.51

		UNIT	COST
06999.20	**FINISH CARPENTRY, Cont'd...**		
1940	Red Cedar Hand Splits - #1, 24" - 1/2" to 3/4"	S.F.	3.99
1950	24" - 3/4" to 1 1/4"	"	4.45
1960	#2	"	3.99
1970	#3	"	3.75
2000	Add for Fire Retardant	"	0.55
2010	Add for 3/8" Backer Board	"	0.93
2020	Add for Metal Corners	EA.	1.80
2030	Add for Ridges, Hips and Corners	L.F.	4.92
2040	Add for 8" to Weather	PCT.	25.00
2100	Facia - Pine #2 1" x 8"	L.F.	2.29
2110	Cedar #3 1" x 8"	"	3.64
2120	Redwood - Clear 1" x 8"	"	4.90
2130	Plywood 5/8" - ACX 1" x 8"	"	2.38
2140	Facia - Pine #2 1" x 8"	S.F.	3.46
2200	Cedar #3 1" x 8"	"	5.14
2210	Redwood - Clear 1" x 8"	"	7.46
2220	Plywood 5/8" - ACX 1" x 8"	"	3.53
2300	FINISH WALLS (Interior) (15% Added for Waste)	"	3.89
2400	Boards, Cedar - #3 1" x 6" and 1" x 8"	"	3.89
2410	Knotty 1" x 6" and 1" x 8"	"	4.10
2420	D Grade 1" x 6" and 1" x 8"	"	4.98
2430	Aromatic 1" x 6" and 1" x 8"	"	5.14
2440	Redwood - Construction 1" x 6" and 1" x 8"	"	6.18
2450	Clear 1" x 6" and 1" x 8"	"	6.40
2460	Fir - Beaded 5/8" x 4"	"	5.48
2470	Pine - #2 1" x 6" and 1" x 8"	"	6.93
2500	Hardboard (Paneling) Tempered - 1/8"	"	5.05
2510	1/4"	"	3.83
2600	Plywood (Prefinished Paneling) 1/4" Birch - Natural	"	1.56
2610	3/4" Birch - Natural	"	1.73
2620	1/4" Birch - White	"	3.34
2630	1/4" Oak - Rotary Cut	"	4.43
2640	3/4"	"	5.80
2700	1/4" Oak - White	"	4.25
2710	1/4" Mahogany (Lauan)	"	5.31
2720	3/4"	"	7.07
2730	3/4" Mahogany (African)	"	3.33
2740	1/4" Walnut	"	4.78
2800	Gypsum Board (Paneling)	"	5.85
2810	Prefinished	"	6.39
2820	Red Oak Plastic		
06999.30	**MILLWORK & CUSTOM WOODWORK**		
1000	CUSTOM CABINET WORK (Red Oak or Birch)		
1010	Base Cabinets - Avg. 35" H x 24" D with Drawer	L.F.	330
1020	Sink Fronts	"	133
1030	Corner Cabinets	"	280
1040	Add per Drawer	"	57.00
1050	Upper Cabinets - Avg. 30" High x 12" Deep	"	174
2000	Utility Cabinets - Avg. 84" High x 24" Deep	"	410
2010	China or Corner Cabinets - 84" High	EA.	790
2020	Oven Cabinets - 84" High x 24" Deep	"	460

		UNIT	COST
06999.30	**MILLWORK & CUSTOM WOODWORK, Cont'd...**		
2030	Vanity Cabinets - Avg. 30" High x 21" Deep	EA.	280
2040	Deduct for Prefinishing wood	PCT.	10.00
2050	Base Cabinets - Avg. 35" H x 24" D with Drawer	EA.	207
2060	Sink Fronts	L.F.	132
2070	Corner Cabinets	"	300
2080	Add per Drawer	"	58.25
2090	Upper Cabinets - Avg. 30" High x 12" Deep	"	171
3000	Utility Cabinets - Avg. 84" High x 24" Deep	"	400
3010 .	China or Corner Cabinets - 84" High	"	800
3020	Oven Cabinets - 84" High x 24" Deep	"	450
3030	Vanity Cabinets - Avg. 30" High x 21" Deep	"	270
4000	COUNTER TOPS - 25"		
4010	Plastic Laminated with 4" Back Splash	L.F.	78.50
4020	Deduct for No Back Splash	"	12.81
4030	Granite - 1 1/4" - Artificial	"	140
4040	3/4" - Artificial	"	111
4050	Marble	"	135
4070	Wood Cutting Block	"	135
5000	Stainless Steel	"	193
5010	Polyester-Acrylic Solid Surface	"	220
5020	Quartz	"	122
5030	Plastic (Polymer)	"	83.00
6000	CUSTOM DOOR FRAMES - Including 2 Sides Trim		
6010	Birch	EA.	360
6020	Fir	"	220
6030	Poplar	"	220
6040	Oak	"	410
6050	Pine	"	300
6060	Walnut	"	480
8000	MOULDINGS AND TRIM		
8010	Apron 7/16" x 2"	L.F.	2.84
8020	Astragal 1 3/4" x 2 1/4"	"	6.58
8030	Base 7/16" x 2 3/4"	"	4.09
8050	Base Shoe 7/16" x 2 3/4"	"	3.26
8060	Batten Strip 5/8" x 1 5/8"	"	2.83
8070	Brick Mould 1 1/14" x 2"	"	3.27
9080	Casing 11/16" x 2 1/4"	"	3.55
9100	Chair Rail 5/8" x 1 3/4"	"	3.73
9110	Closet Rod 1 5/16"	"	3.89
9120	Corner Bead 1 1/8" x 1 1/8"	"	4.81
9130	Cove Moulding 3/4" x 3/4"	"	3.06
9140	Crown Moulding 9/16" x 3 5/8"	"	4.57
9160	Drip Cap	"	3.67
9170	Half Round 1/2" x 1"	"	3.93
9180	Hand Rail 1 5/8" x 1 3/4"	"	5.53
9190	Hook Strip 5/8" x 2 1/2"	"	3.15
9200	Picture Mould 3/4" x 1 1/2"	"	3.51
9210	Quarter Round 3/4" x 3/4"	"	2.48
9220	1/2" x 1/2"	"	1.89
9230	Sill 3/4" x 2 1/2"	"	4.19
9240	Stool 11/16 x 2 1/2"	"	5.06
9250	BIRCH		

		UNIT	COST
06999.30	**MILLWORK & CUSTOM WOODWORK, Cont'd...**		
9260	STAIRS (Treads, Risers, Skirt Boards)	L.F.	46.50
9270	SHELVING (12" Deep)	S.F.	49.25
9280	CUSTOM PANELING	"	33.25
9290	OAK		
9300	STAIRS (Treads, Risers, Skirt Boards)	L.F.	36.50
9310	SHELVING (12" Deep)	S.F.	44.00
9320	CUSTOM PANELING	"	25.00
9330	THRESHOLDS - 3/4" x 3 1/2"	L.F.	24.00
9340	PINE		
9350	STAIRS (Treads, Risers, Skirt Boards)	L.F.	27.25
9360	SHELVING (12" Deep)	"	30.50
9370	CUSTOM PANELING	S.F.	20.82
06999.40	**GLUE LAMINATE**		
0010	Arches 80' Span	S.F.	11.98
0020	Beams & Purlins 40' Span	"	8.21
0030	Deck Fir - 3" x 6"	"	9.68
0040	4" x 6"	"	10.76
0050	Deck Cedar - 3" x 6"	"	12.94
0060	4" x 6"	"	14.40
0070	Deck Pine - 3" x 6"	"	8.81
06999.50	**PREFABRICATED WOOD COMPONENTS**		
1000	WOOD TRUSSED RAFTERS		
1010	24" O.C. - 4 -12 Pitch		
2000	Span To 16' w/ Supports	EA.	164
2010	20'	"	168
2030	24'	"	198
2050	28'	"	210
2070	32'	"	260
3000	Add for 5-12 Pitch	PCT.	5.00
3010	Add for 6-12 Pitch	"	10.00
3020	Add for Scissor Truss	EA.	35.75
3030	Add for Gable End 24'	"	77.50
3040	Span To 16' w/ Supports	S.F.	5.09
3050	20'	"	4.11
3070	24'	"	3.89
3100	28'	"	3.85
3120	32'	"	4.39
5000	Add for 5-12 Pitch	PCT.	15.00
5010	Add for 6-12 Pitch	"	25.00
5020	Add for Scissor Truss	S.F.	28.50
5030	Add for Gable End 24'	"	0.50
6000	WOOD FLOOR TRUSS JOISTS (TJI)		
6010	Open Web - Wood or Metal - 24" O.C.		
6020	Up To 23' Span x 12" Single	L.F.	10.23
6030	To 24' Span x 15" Cord	"	10.49
6050	To 30' Span x 21"	"	11.43
6060	To 25' Span x 15" Double	"	9.97
6070	To 30' Span x 18" Cord	"	10.54
6080	To 36' Span x 21"	"	11.75
7000	Plywood Web - 24" O.C.		
7010	Up To 15' x 9 1/2"	L.F.	6.62

		UNIT	COST
06999.50	**PREFABRICATED WOOD COMPONENTS, Cont'd...**		
7030	To 21' x 14"	L.F.	7.37
7040	To 22' x 16"	"	8.32
8000	Open Web - Wood or Metal - 24" O.C.		
8010	Up To 23' Span x 12" Single	S.F.	5.06
8020	To 24' Span x 15" Cord	"	5.24
8030	To 27' Span x 18"	"	5.37
8040	To 30' Span x 21"	"	5.69
8050	To 25' Span x 15" Double	"	4.98
8060	To 30' Span x 18" Cord	"	5.09
8070	To 36' Span x 21"	"	5.76
8500	Plywood Web - 24" O.C.		
8510	Up To 15' x 9 1/2"	S.F.	3.32
8520	To 19' x 11 7/8"	"	3.32
8530	To 21' x 14"	"	3.72
8540	To 22' x 16"	"	4.11
06999.51	**LAMINATED VENEER STRUCTURAL BEAMS**		
0005	Micro Lam - Plywood - 24" O.C.		
0010	9 1/2" x 1 3/4"	L.F.	9.90
0020	11 7/8" x 1 3/4"	"	10.74
0030	14" x 1 3/4"	"	12.72
0040	16" x 1 3/4"	"	14.07
1000	Glue Lam- Dimension Lumber- 24" O.C.		
1010	9" x 3 1/2"	L.F.	18.66
1020	12" x 3 1/2"	"	23.00
1040	9" x 5 1/2"	"	27.25
1050	12" x 5 1/2"	"	33.00
1070	18" x 5 1/2"	"	51.25
2000	Add for Architectural Grade	PCT.	25.00
2010	Add for 6 3/4"	"	40.00
3000	Micro Lam - Plywood - 24" O.C.		
3010	9 1/2" x 1 3/4"	S.F.	4.97
3020	11 7/8" x 1 3/4"	"	5.36
3040	16" x 1 3/4"	"	6.95
4000	Glue Lam - Dimension Lumber - 24" O.C.		
4010	9" x 3 1/2"	S.F.	9.26
4020	12" x 3 1/2"	"	11.45
4040	9" x 5 1/2"	"	13.61
4050	12" x 5 1/2"	"	16.63
4070	18" x 5 1/2"	"	25.50
06999.60	**WOOD TREATMENTS**		
0010	PRESERVATIVES – PRESSURE TREATED		
0020	Dimensions	B.F.	0.28
0030	Timbers	"	0.41
0040	FIRE RETARDENTS		
0050	Dimensions & Timbers	B.F.	0.53
06999.70	**ROUGH HARDWARE**		
1000	NAILS	B.F.	0.02
2000	JOIST HANGERS	EA.	3.49
3000	BOLTS - 5/8" x 12"	"	4.59

Design & Construction Resources

TABLE OF CONTENTS   PAGE

		UNIT	LABOR	MAT.	TOTAL
07100.10	**WATERPROOFING**				
0100	Membrane waterproofing, elastomeric				
1020	Butyl				
1040	1/32" thick	S.F.	1.62	1.05	2.67
1060	1/16" thick	"	1.69	1.37	3.06
1140	Neoprene				
1160	1/32" thick	S.F.	1.62	1.80	3.42
1180	1/16" thick	"	1.69	2.94	4.63
1260	Plastic vapor barrier (polyethylene)				
1280	4 mil	S.F.	0.16	0.04	0.20
1300	6 mil	"	0.16	0.06	0.22
1320	10 mil	"	0.20	0.09	0.29
1400	Bituminous membrane waterproofing, asphalt felt, 15 lb.				
1440	One ply	S.F.	1.01	0.62	1.63
1460	Two ply	"	1.23	0.73	1.96
1480	Three ply	"	1.45	0.91	2.36
07160.10	**BITUMINOUS DAMPPROOFING**				
0100	Building paper, asphalt felt				
0120	15 lb	S.F.	1.62	0.15	1.77
0140	30 lb	"	1.69	0.28	1.97
1000	Asphalt dampproofing, troweled, cold, primer plus				
1020	1 coat	S.F.	1.35	0.68	2.03
1040	2 coats	"	2.03	1.43	3.46
1060	3 coats	"	2.53	2.04	4.57
1200	Fibrous asphalt dampproofing, hot troweled, primer plus				
1220	1 coat	S.F.	1.62	0.68	2.30
1240	2 coats	"	2.25	1.43	3.68
1260	3 coats	"	2.90	2.04	4.94
07190.10	**VAPOR BARRIERS**				
0980	Vapor barrier, polyethylene				
1000	2 mil	S.F.	0.20	0.01	0.21
1010	6 mil	"	0.20	0.05	0.25
1020	8 mil	"	0.22	0.06	0.28
1040	10 mil	"	0.22	0.07	0.29
07210.10	**BATT INSULATION**				
0980	Ceiling, fiberglass, unfaced				
1000	3-1/2" thick, R11	S.F.	0.47	0.38	0.85
1020	6" thick, R19	"	0.54	0.50	1.04
1030	9" thick, R30	"	0.62	1.00	1.62
1035	Suspended ceiling, unfaced				
1040	3-1/2" thick, R11	S.F.	0.45	0.38	0.83
1060	6" thick, R19	"	0.50	0.50	1.00
1070	9" thick, R30	"	0.58	1.00	1.58
1075	Crawl space, unfaced				
1080	3-1/2" thick, R11	S.F.	0.62	0.38	1.00
1100	6" thick, R19	"	0.67	0.50	1.17
1120	9" thick, R30	"	0.73	1.00	1.73
2000	Wall, fiberglass				
2010	Paper backed				
2020	2" thick, R7	S.F.	0.42	0.25	0.67
2040	3" thick, R8	"	0.45	0.27	0.72
2060	4" thick, R11	"	0.47	0.45	0.92

		UNIT	LABOR	MAT.	TOTAL
07210.10	**BATT INSULATION, Cont'd...**				
2080	6" thick, R19	S.F.	0.50	0.67	1.17
2090	Foil backed, 1 side				
2100	2" thick, R7	S.F.	0.42	0.58	1.00
2120	3" thick, R11	"	0.45	0.61	1.06
2140	4" thick, R14	"	0.47	0.64	1.11
2160	6" thick, R21	"	0.50	0.84	1.34
07210.20	**BOARD INSULATION**				
2200	Perlite board, roof				
2220	1.00" thick, R2.78	S.F.	0.33	0.58	0.91
2240	1.50" thick, R4.17	"	0.35	0.90	1.25
2580	Rigid urethane				
2600	1" thick, R6.67	S.F.	0.33	1.03	1.36
2640	1.50" thick, R11.11	"	0.35	1.40	1.75
2780	Polystyrene				
2800	1.0" thick, R4.17	S.F.	0.33	0.41	0.74
2820	1.5" thick, R6.26	"	0.35	0.63	0.98
07210.60	**LOOSE FILL INSULATION**				
1000	Blown-in type				
1010	Fiberglass				
1020	5" thick, R11	S.F.	0.33	0.36	0.69
1040	6" thick, R13	"	0.40	0.41	0.81
1060	9" thick, R19	"	0.58	0.50	1.08
07210.70	**SPRAYED INSULATION**				
1000	Foam, sprayed on				
1010	Polystyrene				
1020	1" thick, R4	S.F.	0.40	0.61	1.01
1040	2" thick, R8	"	0.54	1.19	1.73
1050	Urethane				
1060	1" thick, R4	S.F.	0.40	0.58	0.98
1080	2" thick, R8	"	0.54	1.11	1.65
07310.10	**ASPHALT SHINGLES**				
1000	Standard asphalt shingles, strip shingles				
1020	210 lb/square	SQ.	50.00	68.00	118
1040	235 lb/square	"	56.00	72.00	128
1060	240 lb/square	"	63.00	75.00	138
1080	260 lb/square	"	72.00	110	182
1100	300 lb/square	"	84.00	120	204
1120	385 lb/square	"	100	160	260
5980	Roll roofing, mineral surface				
6000	90 lb	SQ.	35.75	41.75	77.50
6020	110 lb	"	41.75	69.00	111
6040	140 lb	"	50.00	72.00	122
07310.50	**METAL SHINGLES**				
0980	Aluminum, .020" thick				
1000	Plain	SQ.	100	260	360
1020	Colors	"	100	280	380
1960	Steel, galvanized				
1980	26 ga.				
2000	Plain	SQ.	100	260	360
2020	Colors	"	100	340	440
2030	24 ga.				

		UNIT	LABOR	MAT.	TOTAL
07310.50	**METAL SHINGLES, Cont'd...**				
2040	Plain	SQ.	100	280	380
2060	Colors	"	100	350	450
07310.60	**SLATE SHINGLES**				
0960	Slate shingles				
0980	Pennsylvania				
1000	Ribbon	SQ.	250	630	880
1020	Clear	"	250	810	1,060
1030	Vermont				
1040	Black	SQ.	250	680	930
1060	Gray	"	250	750	1,000
1070	Green	"	250	760	1,010
1080	Red	"	250	1,380	1,630
07310.70	**WOOD SHINGLES**				
1000	Wood shingles, on roofs				
1010	White cedar, #1 shingles				
1020	4" exposure	SQ.	170	230	400
1040	5" exposure	"	130	210	340
1050	#2 shingles				
1060	4" exposure	SQ.	170	160	330
1080	5" exposure	"	130	140	270
1090	Resquared and rebutted				
1100	4" exposure	SQ.	170	210	380
1120	5" exposure	"	130	170	300
1140	On walls				
1150	White cedar, #1 shingles				
1160	4" exposure	SQ.	250	230	480
1180	5" exposure	"	200	210	410
1200	6" exposure	"	170	170	340
1210	#2 shingles				
1220	4" exposure	SQ.	250	160	410
1240	5" exposure	"	200	140	340
1260	6" exposure	"	170	120	290
1300	Add for fire retarding	"			110
07310.80	**WOOD SHAKES**				
2010	Shakes, hand split, 24" red cedar, on roofs				
2020	5" exposure	SQ.	250	260	510
2040	7" exposure	"	200	250	450
2060	9" exposure	"	170	230	400
2080	On walls				
2100	6" exposure	SQ.	250	250	500
2120	8" exposure	"	200	240	440
2140	10" exposure	"	170	220	390
3000	Add for fire retarding	"			75.00
07310.90	**CLAY TILE ROOF (assorted colors)**				
1010	Spanish tile, terracota	SQ.	130	390	520
1020	Mission tile, terracota	"	170	820	990
1030	French tile, rustic blue, green patina	"	140	950	1,090
07460.10	**METAL SIDING PANELS**				
1000	Aluminum siding panels				
1020	Corrugated				
1030	Plain finish				

		UNIT	LABOR	MAT.	TOTAL
07460.10	**METAL SIDING PANELS, Cont'd...**				
1040	.024"	S.F.	2.32	1.77	4.09
1060	.032"	"	2.32	2.09	4.41
1070	Painted finish				
1080	.024"	S.F.	2.32	2.21	4.53
1100	.032"	"	2.32	2.53	4.85
2000	Steel siding panels				
2040	Corrugated				
2080	22 ga.	S.F.	3.87	2.25	6.12
2100	24 ga.	"	3.87	2.05	5.92
07460.50	**PLASTIC SIDING**				
1000	Horizontal vinyl siding, solid				
1010	8" wide				
1020	Standard	S.F.	2.00	1.23	3.23
1040	Insulated	"	2.00	1.49	3.49
1050	10" wide				
1060	Standard	S.F.	1.86	1.27	3.13
1080	Insulated	"	1.86	1.52	3.38
8500	Vinyl moldings for doors and windows	L.F.	2.08	0.79	2.87
07460.60	**PLYWOOD SIDING**				
1000	Rough sawn cedar, 3/8" thick	S.F.	1.74	1.79	3.53
1020	Fir, 3/8" thick	"	1.74	0.99	2.73
1980	Texture 1-11, 5/8" thick				
2000	Cedar	S.F.	1.86	2.42	4.28
2020	Fir	"	1.86	1.69	3.55
2040	Redwood	"	1.79	2.59	4.38
2060	Southern Yellow Pine	"	1.86	1.37	3.23
07460.80	**WOOD SIDING**				
1000	Beveled siding, cedar				
1010	A grade				
1040	1/2 x 8	S.F.	2.08	4.42	6.50
1060	3/4 x 10	"	1.74	5.68	7.42
1070	Clear				
1100	1/2 x 8	S.F.	2.08	4.92	7.00
1120	3/4 x 10	"	1.74	6.60	8.34
1130	B grade				
1160	1/2 x 8	S.F.	20.75	5.25	26.00
1180	3/4 x 10	"	1.74	4.95	6.69
2000	Board and batten				
2010	Cedar				
2020	1x6	S.F.	2.61	5.12	7.73
2040	1x8	"	2.08	4.66	6.74
2060	1x10	"	1.86	4.21	6.07
2080	1x12	"	1.68	3.77	5.45
2090	Pine				
2100	1x6	S.F.	2.61	1.29	3.90
2120	1x8	"	2.08	1.26	3.34
2140	1x10	"	1.86	1.21	3.07
2160	1x12	"	1.68	1.12	2.80
3000	Tongue and groove				
3010	Cedar				
3020	1x4	S.F.	2.90	4.80	7.70

		UNIT	LABOR	MAT.	TOTAL
07460.80	**WOOD SIDING, Cont'd...**				
3040	1x6	S.F.	2.74	4.63	7.37
3060	1x8	"	2.61	4.33	6.94
3080	1x10	"	2.48	4.26	6.74
3090	Pine				
3100	1x4	S.F.	2.90	1.44	4.34
3120	1x6	"	2.74	1.36	4.10
3140	1x8	"	2.61	1.27	3.88
3160	1x10	"	2.48	1.21	3.69
07510.10	**BUILT-UP ASPHALT ROOFING**				
0980	Built-up roofing, asphalt felt, including gravel				
1000	2 ply	SQ.	130	80.00	210
1500	3 ply	"	170	110	280
2000	4 ply	"	200	160	360
2195	Cant strip, 4" x 4"				
2200	Treated wood	L.F.	1.43	1.92	3.35
2260	Foamglass	"	1.25	1.65	2.90
8000	New gravel for built-up roofing, 400 lb/sq	SQ.	100	32.00	132
07530.10	**SINGLE-PLY ROOFING**				
2000	Elastic sheet roofing				
2060	Neoprene, 1/16" thick	S.F.	0.62	2.83	3.45
2115	PVC				
2120	45 mil	S.F.	0.62	2.03	2.65
2200	Flashing				
2220	Pipe flashing, 90 mil thick				
2260	1" pipe	EA.	12.50	28.50	41.00
2360	Neoprene flashing, 60 mil thick strip				
2380	6" wide	L.F.	4.18	1.73	5.91
2390	12" wide	"	6.27	3.42	9.69
07610.10	**METAL ROOFING**				
1000	Sheet metal roofing, copper, 16 oz, batten seam	SQ.	330	1,200	1,530
1020	Standing seam	"	310	1,170	1,480
2000	Aluminum roofing, natural finish				
2005	Corrugated, on steel frame				
2010	.0175" thick	SQ.	140	120	260
2040	.0215" thick	"	140	160	300
2060	.024" thick	"	140	190	330
2080	.032" thick	"	140	240	380
2100	V-beam, on steel frame				
2120	.032" thick	SQ.	140	250	390
2130	.040" thick	"	140	270	410
2140	.050" thick	"	140	340	480
2200	Ridge cap				
2220	.019" thick	L.F.	1.67	3.85	5.52
2500	Corrugated galvanized steel roofing, on steel frame				
2520	28 ga.	SQ.	140	180	320
2540	26 ga.	"	140	210	350
2550	24 ga.	"	140	240	380
2560	22 ga.	"	140	260	400

		UNIT	LABOR	MAT.	TOTAL
07620.10	**FLASHING AND TRIM**				
0050	Counter flashing				
0060	Aluminum, .032"	S.F.	5.01	1.58	6.59
0100	Stainless steel, .015"	"	5.01	5.06	10.07
0105	Copper				
0110	16 oz.	S.F.	5.01	7.49	12.50
0112	20 oz.	"	5.01	8.89	13.90
0114	24 oz.	"	5.01	10.75	15.76
0116	32 oz.	"	5.01	13.25	18.26
0118	Valley flashing				
0120	Aluminum, .032"	S.F.	3.13	1.58	4.71
0130	Stainless steel, .015	"	3.13	5.06	8.19
0135	Copper				
0140	16 oz.	S.F.	3.13	7.49	10.62
0160	20 oz.	"	4.18	8.89	13.07
0180	24 oz.	"	3.13	10.75	13.88
0200	32 oz.	"	3.13	13.25	16.38
0380	Base flashing				
0400	Aluminum, .040"	S.F.	4.18	2.60	6.78
0410	Stainless steel, .018"	"	4.18	6.05	10.23
0415	Copper				
0420	16 oz.	S.F.	4.18	7.49	11.67
0422	20 oz.	"	3.13	8.89	12.02
0424	24 oz.	"	4.18	10.75	14.93
0426	32 oz.	"	4.18	13.25	17.43
07620.20	**GUTTERS AND DOWNSPOUTS**				
1500	Copper gutter and downspout				
1520	Downspouts, 16 oz. copper				
1530	Round				
1540	3" dia.	L.F.	3.34	10.50	13.84
1550	4" dia.	"	3.34	13.00	16.34
1800	Gutters, 16 oz. copper				
1810	Half round				
1820	4" wide	L.F.	5.01	9.38	14.39
1840	5" wide	"	5.57	11.50	17.07
1860	Type K				
1880	4" wide	L.F.	5.01	10.50	15.51
1890	5" wide	"	5.57	11.00	16.57
3000	Aluminum gutter and downspout				
3005	Downspouts				
3010	2" x 3"	L.F.	3.34	1.09	4.43
3030	3" x 4"	"	3.58	1.50	5.08
3035	4" x 5"	"	3.85	1.67	5.52
3038	Round				
3040	3" dia.	L.F.	3.34	1.84	5.18
3050	4" dia.	"	3.58	2.36	5.94
3240	Gutters, stock units				
3260	4" wide	L.F.	5.28	1.79	7.07
3270	5" wide	"	5.57	2.13	7.70
4101	Galvanized steel gutter and downspout				
4111	Downspouts, round corrugated				
4121	3" dia.	L.F.	3.34	1.84	5.18

		UNIT	LABOR	MAT.	TOTAL
07620.20	**GUTTERS AND DOWNSPOUTS, Cont'd...**				
4131	4" dia.	L.F.	3.34	2.48	5.82
4141	5" dia.	"	3.58	3.69	7.27
4151	6" dia.	"	3.58	4.90	8.48
4161	Rectangular				
4171	2" x 3"	L.F.	3.34	1.67	5.01
4191	3" x 4"	"	3.13	2.39	5.52
4201	4" x 4"	"	3.13	3.00	6.13
4300	Gutters, stock units				
4310	5" wide				
4320	Plain	L.F.	5.57	1.61	7.18
4330	Painted	"	5.57	1.76	7.33
4335	6" wide				
4340	Plain	L.F.	5.90	2.25	8.15
4360	Painted	"	5.90	2.53	8.43
07810.10	**PLASTIC SKYLIGHTS**				
1030	Single thickness, not including mounting curb				
1040	2' x 4'	EA.	63.00	360	423
1050	4' x 4'	"	84.00	490	574
1060	5' x 5'	"	130	650	780
1070	6' x 8'	"	170	1,390	1,560
07920.10	**CAULKING**				
0100	Caulk exterior, two component				
0120	1/4 x 1/2	L.F.	2.61	0.39	3.00
0140	3/8 x 1/2	"	2.90	0.60	3.50
0160	1/2 x 1/2	"	3.26	0.82	4.08
0220	Caulk interior, single component				
0240	1/4 x 1/2	L.F.	2.48	0.26	2.74
0260	3/8 x 1/2	"	2.74	0.37	3.11
0280	1/2 x 1/2	"	3.07	0.49	3.56

		UNIT	COST
07999.10	**WATERPROOFING**		
0010	1-Ply Membrane Felt 15#	S.F.	1.71
0020	2-Ply Membrane Felt 15#	"	2.60
0030	Hydrolithic	"	2.81
0040	Elastomeric Rubberized Asphalt with Poly Sheet	"	2.60
0060	Metallic Oxide 3-Coat	"	4.44
0070	Vinyl Plastic	"	3.10
0080	Bentonite 3/8" - Trowel	"	2.64
0090	5/8" - Panels	"	3.00
07999.20	**DAMPPROOFING**		
0010	Asphalt Trowel Mastic 1/16"	S.F.	1.05
0020	1/8"	"	1.41
0030	Spray Liquid 1-Coat	"	0.65
0040	2-Coat	"	0.91
0050	Brush Liquid 2-Coat	"	1.02
1000	Hot Mop 1-Coat and Primer	"	1.71
1010	1 Fibrous Asphalt	"	1.87
1020	Cementitious Per Coat 1/2" Coat	"	1.10
1030	Silicone 1-Coat	"	0.71
1040	2-Coat	"	1.27
1050	Add for Scaffold and Lift Operations	"	0.65
07999.30	**BUILDING INSULATION**		
1000	FLEXIBLE		
1010	Fiberglass 2 1/4" R 7.40	S.F.	0.76
1020	3 1/2" R 11.00	"	0.89
1040	6" R 19.00	"	1.12
1060	12" R 38.00	"	1.55
1100	Add for Ceiling Work	"	0.10
1110	Add for Paper Faced	"	0.12
1120	Add for Polystyrene Barrier (2m)	"	0.12
1130	Add for Scaffold Work	"	0.76
1200	RIGID		
1300	Fiberglass 1" R 4.35 3# Density	S.F.	1.18
1310	1 1/2" R 6.52 3# Density	"	1.52
1320	2" R 8.70 3# Density	"	1.79
1400	Styrene, Molded 1" R 4.3.5	"	5.64
1410	1 1/2" R 6.52	"	1.00
1420	2" R 7.69	"	1.35
1430	2" T&G R 7.69	"	1.73
1500	Styrene, Extruded 1" R 5.40	"	1.23
1510	1 1/2" R 6.52	"	1.45
1520	2" R 7.69	"	1.55
1530	2" T&G R 7.69	"	1.73
1600	Perlite 1" R 2.78	"	1.23
1610	2" R 5.56	"	1.96
1700	Urethane 1" R 6.67	"	1.45
1710	2" R 13.34	"	1.85
1720	Add for Glued Applications	"	0.10
1800	LOOSE 1" Styrene R 3.8	"	0.86
1810	1" Fiberglass R 2.2	"	0.83
1820	1" Rock Wool R 2.9	"	0.85
1830	1" Cellulose R 3.7	"	0.81

		UNIT	COST
07999.30	**BUILDING INSULATION, Cont'd...**		
1900	FOAMED Urethane Per Inch	S.F.	2.28
2000	SPRAYED Cellulose Per Inch	"	1.53
2100	Polystyrene	"	1.83
2200	Urethane	"	2.59
2300	ALUMINUM PAPER	"	0.69
2400	VINYL FACED FIBERGLASS	"	2.07
07999.80	**SHINGLE ROOFING**		
1000	ASPHALT SHINGLES 235# Seal Down	S.F.	1.37
1010	300# Laminated	"	1.56
1020	325# Fire Resistant	"	2.24
1030	Timberline	"	2.20
1040	340# Tab Lock	"	2.59
2000	ASPHALT ROLL 90#	"	0.75
3000	FIBERGLASS SHINGLES 215#	"	1.33
3010	250#	"	1.50
4000	RED CEDAR 16" x 5" to Weather #1 Grade	"	4.00
4010	#2 Grade	"	3.55
4020	#3 Grade	"	2.54
4030	24" x 10" to Weather - Hand Splits - 1/2" x 3/4" #	"	12.75
4040	3/4" x 1 1/4" #	"	12.86
4050	Add for Fire Retardant	"	1.00
5000	METAL Aluminum 020 mil	"	4.30
6000	Anodized 020 Mil	"	4.99
7000	Steel, Enameled Colored	"	6.42
8000	Galvanized Colored	"	5.55
9000	Add to Above for 15# Felt Underlayment		
9010	Add for Base Starter	S.F.	0.78
9020	Add for Ice & Water Starter	"	1.30
9030	Add for Boston Ridge	"	2.94
9040	Add for Removal & Haul Away	"	1.12
9050	Add for Pitches Over 5-12 each Pitch Increase	"	0.06
9060	Add for Chimneys, Skylights and Bay Windows	EA.	84.50
07999.90	**BUILT-UP ROOFING**		
1000	MEMBRANE		
1010	3-Ply Asphalt & Gravel - R 10.0	S.F.	710
1020	R 16.6	"	730
1030	4-Ply - R 10.0	"	730
1040	R 16.6	"	750
1050	5-Ply - R 10.0	"	770
1060	R 16.6	"	800
2000	Add for Sheet Rock over Steel Deck - 5/8"	"	152
2010	Add for Fiberglass Insulation	"	51.25
2020	Add for Pitch and Gravel	"	78.75
2030	Add for Sloped Roofs	"	55.50
2040	Add for Thermal Barrier	"	42.75
2050	Add for Upside Down Roofing System	"	152
3000	SINGLE-PLY - 60M Butylene Roofing & Gravel Ballast - R 10.0	"	540
3010	Mech. Fastened - R 10.0	"	520
3020	PVC & EPDM Roofing & Gravel Ballast - R 10.0	"	500
3030	Mech. Fastened - R 10.0	"	520
3040	Blocking and Cants Not Included		

TABLE OF CONTENTS   PAGE

08110.10 METAL DOORS

		UNIT	LABOR	MAT.	TOTAL
1000	Flush hollow metal, std. duty, 20 ga., 1-3/8" thick				
1020	2-6 x 6-8	EA.	58.00	300	358
1080	3-0 x 6-8	"	58.00	360	418
1090	1-3/4" thick				
1100	2-6 x 6-8	EA.	58.00	360	418
1150	3-0 x 6-8	"	58.00	410	468
1200	2-6 x 7-0	"	58.00	390	448
1240	3-0 x 7-0	"	58.00	430	488
2110	Heavy duty, 20 ga., unrated, 1-3/4"				
2130	2-8 x 6-8	EA.	58.00	390	448
2135	3-0 x 6-8	"	58.00	420	478
2140	2-8 x 7-0	"	58.00	450	508
2150	3-0 x 7-0	"	58.00	430	488
2200	18 ga., 1-3/4", unrated door				
2210	2-0 x 7-0	EA.	58.00	420	478
2235	2-6 x 7-0	"	58.00	420	478
2260	3-0 x 7-0	"	58.00	470	528
2270	3-4 x 7-0	"	58.00	480	538
2310	2", unrated door				
2320	2-0 x 7-0	EA.	65.00	420	485
2340	2-6 x 7-0	"	65.00	420	485
2360	3-0 x 7-0	"	65.00	470	535
2370	3-4 x 7-0	"	65.00	480	545
2400	Galvanized metal door				
2410	3-0 x 7-0	EA.	65.00	500	565
2450	For lead lining in doors	"			970
2460	For sound attenuation	"			88.00
4280	Vision glass				
4300	8" x 8"	EA.	65.00	100	165
4320	8" x 48"	"	65.00	150	215
4340	Fixed metal louver	"	52.00	200	252
4350	For fire rating, add				
4370	3 hr door	EA.			390
4380	1-1/2 hr door	"			160
4400	3/4 hr door	"			81.00
4430	1' extra height, add to material, 20%				
4440	1'6" extra height, add to material, 60%				
4470	For dutch doors with shelf, add to material, 100%				
5000	Stainless steel, general application				
5010	3'x7'	EA.	580	1,620	2,200
5020	5'X7'	"	770	2,690	3,460
6000	Heavy impact, s-core, 18 ga., stainless				
6010	3'x7'	EA.	580	1,960	2,540
6020	5'X7'	"	770	3,270	4,040

08110.40 METAL DOOR FRAMES

		UNIT	LABOR	MAT.	TOTAL
1000	Hollow metal, stock, 18 ga., 4-3/4" x 1-3/4"				
1020	2-0 x 7-0	EA.	65.00	140	205
1060	2-6 x 7-0	"	65.00	150	215
1100	3-0 x 7-0	"	65.00	160	225
1120	4-0 x 7-0	"	87.00	170	257
1140	5-0 x 7-0	"	87.00	180	267

		UNIT	LABOR	MAT.	TOTAL
08110.40	**METAL DOOR FRAMES, Cont'd...**				
1160	6-0 x 7-0	EA.	87.00	210	297
1500	16 ga., 6-3/4" x 1-3/4"				
1520	2-0 x 7-0	EA.	73.00	160	233
1535	2-6 x 7-0	"	73.00	150	223
1550	3-0 x 7-0	"	73.00	160	233
1560	4-0 x 7-0	"	97.00	190	287
1580	6-0 x 7-0	"	97.00	220	317
08210.10	**WOOD DOORS**				
0980	Solid core, 1-3/8" thick				
1000	Birch faced				
1020	2-4 x 7-0	EA.	65.00	150	215
1060	3-0 x 7-0	"	65.00	160	225
1070	3-4 x 7-0	"	65.00	310	375
1080	2-4 x 6-8	"	65.00	150	215
1090	2-6 x 6-8	"	65.00	150	215
1100	3-0 x 6-8	"	65.00	160	225
1120	Lauan faced				
1140	2-4 x 6-8	EA.	65.00	140	205
1180	3-0 x 6-8	"	65.00	150	215
1200	3-4 x 6-8	"	65.00	160	225
1300	Tempered hardboard faced				
1320	2-4 x 7-0	EA.	65.00	170	235
1360	3-0 x 7-0	"	65.00	210	275
1380	3-4 x 7-0	"	65.00	210	275
1420	Hollow core, 1-3/8" thick				
1440	Birch faced				
1460	2-4 x 7-0	EA.	65.00	140	205
1500	3-0 x 7-0	"	65.00	150	215
1520	3-4 x 7-0	"	65.00	160	225
1600	Lauan faced				
1620	2-4 x 6-8	EA.	65.00	58.00	123
1630	2-6 x 6-8	"	65.00	63.00	128
1660	3-0 x 6-8	"	65.00	82.00	147
1680	3-4 x 6-8	"	65.00	91.00	156
1740	Tempered hardboard faced				
1770	2-6 x 7-0	EA.	65.00	77.00	142
1800	3-0 x 7-0	"	65.00	90.00	155
1820	3-4 x 7-0	"	65.00	98.00	163
1900	Solid core, 1-3/4" thick				
1920	Birch faced				
1940	2-4 x 7-0	EA.	65.00	230	295
1950	2-6 x 7-0	"	65.00	240	305
1970	3-0 x 7-0	"	65.00	230	295
1980	3-4 x 7-0	"	65.00	240	305
2000	Lauan faced				
2020	2-4 x 7-0	EA.	65.00	160	225
2030	2-6 x 7-0	"	65.00	180	245
2060	3-4 x 7-0	"	65.00	200	265
2080	3-0 x 7-0	"	65.00	220	285
2140	Tempered hardboard faced				
2170	2-6 x 7-0	EA.	65.00	230	295

08210.10 WOOD DOORS, Cont'd...		UNIT	LABOR	MAT.	TOTAL
2190	3-0 x 7-0	EA.	65.00	270	335
2200	3-4 x 7-0	"	65.00	290	355
2250	Hollow core, 1-3/4" thick				
2270	Birch faced				
2295	2-6 x 7-0	EA.	65.00	160	225
2320	3-0 x 7-0	"	65.00	170	235
2340	3-4 x 7-0	"	65.00	180	245
2400	Lauan faced				
2430	2-6 x 6-8	EA.	65.00	110	175
2460	3-0 x 6-8	"	65.00	100	165
2480	3-4 x 6-8	"	65.00	110	175
2520	Tempered hardboard				
2550	2-6 x 7-0	EA.	65.00	93.00	158
2580	3-0 x 7-0	"	65.00	100	165
2600	3-4 x 7-0	"	65.00	110	175
2620	Add-on, louver	"	52.00	35.00	87.00
2640	Glass	"	52.00	110	162
2700	Exterior doors, 3-0 x 7-0 x 2-1/2", solid core				
2710	Carved				
2720	One face	EA.	130	1,160	1,290
2740	Two faces	"	130	1,590	1,720
3000	Closet doors, 1-3/4" thick				
3001	Bi-fold or bi-passing, includes frame and trim				
3020	Paneled				
3040	4-0 x 6-8	EA.	87.00	460	547
3060	6-0 x 6-8	"	87.00	520	607
3070	Louvered				
3080	4-0 x 6-8	EA.	87.00	320	407
3100	6-0 x 6-8	"	87.00	380	467
3130	Flush				
3140	4-0 x 6-8	EA.	87.00	230	317
3160	6-0 x 6-8	"	87.00	300	387
3170	Primed				
3180	4-0 x 6-8	EA.	87.00	260	347
3200	6-0 x 6-8	"	87.00	290	377
3210	French Door, Dual-Tempered, Clear-Glass, 6'-8'				
3220	24"x80"x1-3/4"	EA.	100	370	470
3230	with Low-E glass	"	100	500	600
3240	30"x80"x1-3/4"	"	100	420	520
3250	with Low-E glass	"	100	500	600
3260	42"x80"x1-3/4"	"	100	550	650
3270	with Low-E glass	"	100	660	760
3280	24"x96"x1-3/4"	"	100	440	540
3290	with Low-E glass	"	100	550	650
3310	32"x96"x1-3/4"	"	100	490	590
3320	with Low-E glass	"	100	610	710
3330	French door, 10-lite, 1-3/4' thick, 6'-8" high				
3340	24" wide	EA.	100	390	490
3350	30" wide	"	100	480	580
3360	36" wide	"	100	490	590
3370	96" high, 24" wide	"	100	500	600
3380	30" wide	"	100	600	700

		UNIT	LABOR	MAT.	TOTAL
08210.10	**WOOD DOORS, Cont'd...**				
3390	36" wide	EA.	100	780	880
3400	French door, 1-lite, 1-3/4' thick, 6'-8" high				
3410	48" wide	EA.	100	1,380	1,480
3420	56" wide	"	100	1,650	1,750
3430	60" wide	"	100	1,930	2,030
3440	72" wide	"	100	1,810	1,910
3500	Fiberglass, single door, 6'-8" high				
3510	2-4" wide	EA.	65.00	200	265
3520	2-6" wide	"	65.00	300	365
3530	2-8" wide	"	65.00	300	365
3540	3-0" wide	"	65.00	370	435
3550	8'-0" high				
3560	2-4" wide	EA.	65.00	400	465
3570	2-6" wide	"	65.00	500	565
3580	2-8" wide	"	65.00	600	665
3590	3-0" wide	"	65.00	600	665
3600	Fiberglass, double door, 6'-8" high				
3610	56" wide	EA.	100	700	800
3620	60" wide	"	100	800	900
3630	64" wide	"	100	830	930
3640	72" wide	"	100	950	1,050
3650	8'-0" high				
3700	56" wide	EA.	100	260	360
3710	60" wide	"	100	270	370
3720	64" wide	"	100	290	390
3730	72" wide	"	100	310	410
3740	Metal clad, double door, 6'-8" high				
3750	56" wide	EA.	100	280	380
3760	60" wide	"	100	290	390
3770	64" wide	"	100	300	400
3780	72" wide	"	100	310	410
3790	8'-0" high				
3800	56" wide	EA.	100	470	570
3810	60" wide	"	100	500	600
3820	64" wide	"	100	550	650
3830	72" wide	"	100	600	700
3840	For outswinging doors, add min.	"			66.00
3850	Maximum	"			140
08210.90	**WOOD FRAMES**				
0080	Frame, interior, pine				
0100	2-6 x 6-8	EA.	75.00	71.00	146
0160	3-0 x 6-8	"	75.00	79.00	154
0180	5-0 x 6-8	"	75.00	82.00	157
0200	6-0 x 6-8	"	75.00	87.00	162
0220	2-6 x 7-0	"	75.00	81.00	156
0260	3-0 x 7-0	"	75.00	95.00	170
0280	5-0 x 7-0	"	100	100	200
0300	6-0 x 7-0	"	100	110	210
1000	Exterior, custom, with threshold, including trim				
1040	Walnut				
1060	3-0 x 7-0	EA.	130	340	470

		UNIT	LABOR	MAT.	TOTAL
08210.90	**WOOD FRAMES, Cont'd...**				
1080	6-0 x 7-0	EA.	130	390	520
1090	Oak				
1100	3-0 x 7-0	EA.	130	310	440
1120	6-0 x 7-0	"	130	360	490
1200	Pine				
1240	2-6 x 7-0	EA.	100	130	230
1300	3-0 x 7-0	"	100	140	240
1320	3-4 x 7-0	"	100	160	260
1340	6-0 x 7-0	"	170	170	340
3000	Fire-rated wood jambs				
3010	20-Minute, positive or neutral pressure, 3/4"	L.F.	4.35	8.80	13.15
3020	45-60-minute	"	4.35	9.90	14.25
3030	90-minute	"	4.35	11.00	15.35
3040	120-minute	"	4.35	13.25	17.60
08300.10	**SPECIAL DOORS**				
1000	Vault door and frame, class 5, steel	EA.	520	6,550	7,070
1480	Overhead door, coiling insulated				
1500	Chain gear, no frame, 12' x 12'	EA.	650	3,410	4,060
2000	Aluminum, bronze glass panels, 12-9 x 13-0	"	520	3,850	4,370
2200	Garage door, flush insulated metal, primed, 9-0 x 7-0	"	170	1,040	1,210
3000	Sliding metal fire doors, motorized, fusible link, 3 hr.				
3040	3-0 x 6-8	EA.	1,040	3,900	4,940
3060	3-8 x 6-8	"	1,040	3,900	4,940
3080	4-0 x 8-0	"	1,040	4,240	5,280
3100	5-0 x 8-0	"	1,040	4,280	5,320
3200	Metal clad doors, including electric motor				
3210	Light duty				
3220	Minimum	S.F.	8.70	43.00	51.70
3240	Maximum	"	20.75	69.00	89.75
3250	Heavy duty				
3260	Minimum	S.F.	26.00	67.00	93.00
3280	Maximum	"	32.50	110	143
3500	Counter doors, (roll-up shutters), standard, manual				
3510	Opening, 4' high				
3520	4' wide	EA.	430	1,300	1,730
3540	6' wide	"	430	1,760	2,190
3560	8' wide	"	470	1,980	2,450
3580	10' wide	"	650	2,200	2,850
3590	14' wide	"	650	2,750	3,400
3595	6' high				
3600	4' wide	EA.	430	1,540	1,970
3620	6' wide	"	470	2,010	2,480
3630	8' wide	"	520	2,200	2,720
3640	10' wide	"	650	2,480	3,130
3650	14' wide	"	750	2,810	3,560
3660	For stainless steel, add to material, 40%				
3670	For motor operator, add	EA.			1,520
3800	Service doors, (roll up shutters), standard, manual				
3810	Opening				
3820	8' high x 8' wide	EA.	290	1,650	1,940
3840	12' high x 12' wide	"	650	2,310	2,960

		UNIT	LABOR	MAT.	TOTAL
08300.10	**SPECIAL DOORS, Cont'd...**				
3860	16' high x 14' wide	EA.	870	4,240	5,110
3870	20' high x 14' wide	"	1,310	4,790	6,100
3880	24' high x 16' wide	"	1,160	7,810	8,970
3890	For motor operator				
3900	Up to 12-0 x 12-0, add	EA.			1,550
3920	Over 12-0 x 12-0, add	"			1,980
4040	Roll-up doors				
4050	13-0 high x 14-0 wide	EA.	750	1,400	2,150
4060	12-0 high x 14-0 wide	"	750	1,790	2,540
5200	Sectional wood overhead doors, frames not included				
5220	Commercial grade, heavy duty, 1-3/4" thick, manual				
5240	8' x 8'	EA.	430	1,030	1,460
5260	10' x 10'	"	470	1,480	1,950
5280	12' x 12'	"	520	1,930	2,450
5290	Chain hoist				
5300	12' x 16' high	EA.	870	2,860	3,730
5320	14' x 14' high	"	650	3,140	3,790
5340	20' x 8' high	"	1,040	2,690	3,730
5360	16' high	"	1,310	6,050	7,360
5990	Sectional metal overhead doors, complete				
6000	Residential grade, manual				
6020	9' x 7'	EA.	210	710	920
6040	16' x 7'	"	260	1,270	1,530
6100	Commercial grade				
6120	8' x 8'	EA.	430	820	1,250
6140	10' x 10'	"	470	1,090	1,560
6160	12' x 12'	"	520	1,810	2,330
6180	20' x 14', with chain hoist	"	1,040	4,350	5,390
6400	Sliding glass doors				
6420	Tempered plate glass, 1/4" thick				
6430	6' wide				
6440	Economy grade	EA.	170	1,020	1,190
6450	Premium grade	"	170	1,170	1,340
6455	12' wide				
6460	Economy grade	EA.	260	1,430	1,690
6465	Premium grade	"	260	2,140	2,400
6470	Insulating glass, 5/8" thick				
6475	6' wide				
6480	Economy grade	EA.	170	1,250	1,420
6500	Premium grade	"	170	1,610	1,780
6505	12' wide				
6510	Economy grade	EA.	260	1,560	1,820
6515	Premium grade	"	260	2,500	2,760
6520	1" thick				
6525	6' wide				
6530	Economy grade	EA.	170	1,580	1,750
6540	Premium grade	"	170	1,820	1,990
6545	12' wide				
6550	Economy grade	EA.	260	2,450	2,710
6560	Premium grade	"	260	3,580	3,840
6600	Added costs				
6620	Custom quality, add to material, 30%				

		UNIT	LABOR	MAT.	TOTAL
08300.10	**SPECIAL DOORS, Cont'd...**				
6630	Tempered glass, 6' wide, add	S.F.			4.40
6880	Residential storm door				
6900	Minimum	EA.	87.00	160	247
6920	Average	"	87.00	220	307
6940	Maximum	"	130	480	610
08410.10	**STOREFRONTS**				
0135	Storefront, aluminum and glass				
0140	Minimum	S.F.	7.26	24.00	31.26
0150	Average	"	8.30	35.75	44.05
0160	Maximum	"	9.68	72.00	81.68
1020	Entrance doors, premium, incl. glass, closers, panic dev.,etc.				
1030	1/2" thick glass				
1040	3' x 7'	EA.	480	3,190	3,670
1060	6' x 7'	"	730	5,450	6,180
1065	3/4" thick glass				
1070	3' x 7'	EA.	480	3,300	3,780
1080	6' x 7'	"	730	5,500	6,230
1085	1" thick glass				
1090	3' x 7'	EA.	480	3,580	4,060
1100	6' x 7'	"	730	6,330	7,060
1150	Revolving doors				
1151	7'diameter, 7' high				
1160	Minimum	EA.	5,580	24,920	30,500
1170	Average	"	8,920	31,350	40,270
1180	Maximum	"	11,150	40,460	51,610
08510.10	**STEEL WINDOWS**				
0100	Steel windows, primed				
1000	Casements				
1010	Operable				
1020	Minimum	S.F.	3.41	37.00	40.41
1040	Maximum	"	3.87	55.00	58.87
1060	Fixed sash	"	2.90	29.50	32.40
1080	Double hung	"	3.22	55.00	58.22
1100	Industrial windows				
1120	Horizontally pivoted sash	S.F.	3.87	52.00	55.87
1130	Fixed sash	"	3.22	40.75	43.97
1135	Security sash				
1140	Operable	S.F.	3.87	65.00	68.87
1150	Fixed	"	3.22	57.00	60.22
1155	Picture window	"	3.22	27.75	30.97
1160	Projecting sash				
1170	Minimum	S.F.	3.63	48.00	51.63
1180	Maximum	"	3.63	59.00	62.63
1930	Mullions	L.F.	2.90	12.75	15.65
08520.10	**ALUMINUM WINDOWS**				
0110	Jalousie				
0120	3-0 x 4-0	EA.	73.00	320	393
0140	3-0 x 5-0	"	73.00	370	443
0220	Fixed window				
0240	6 sf to 8 sf	S.F.	8.30	15.75	24.05
0250	12 sf to 16 sf	"	6.45	14.00	20.45

		UNIT	LABOR	MAT.	TOTAL
08520.10	**ALUMINUM WINDOWS, Cont'd...**				
0255	Projecting window				
0260	6 sf to 8 sf	S.F.	14.50	35.00	49.50
0270	12 sf to 16 sf	"	9.68	31.50	41.18
0275	Horizontal sliding				
0280	6 sf to 8 sf	S.F.	7.26	22.75	30.01
0290	12 sf to 16 sf	"	5.81	21.00	26.81
1140	Double hung				
1160	6 sf to 8 sf	S.F.	11.50	31.50	43.00
1180	10 sf to 12 sf	"	9.68	28.00	37.68
3010	Storm window, 0.5 cfm, up to				
3020	60 u.i. (united inches)	EA.	29.00	74.00	103
3060	80 u.i.	"	29.00	84.00	113
3080	90 u.i.	"	32.25	86.00	118
3100	100 u.i.	"	32.25	88.00	120
3110	2.0 cfm, up to				
3120	60 u.i.	EA.	29.00	95.00	124
3160	80 u.i.	"	29.00	98.00	127
3180	90 u.i.	"	32.25	110	142
3200	100 u.i.	"	32.25	110	142
08600.10	**WOOD WINDOWS**				
0980	Double hung				
0990	24" x 36"				
1000	Minimum	EA.	52.00	200	252
1002	Average	"	65.00	290	355
1004	Maximum	"	87.00	390	477
1010	24" x 48"				
1020	Minimum	EA.	52.00	230	282
1022	Average	"	65.00	340	405
1024	Maximum	"	87.00	470	557
1030	30" x 48"				
1040	Minimum	EA.	58.00	240	298
1042	Average	"	75.00	340	415
1044	Maximum	"	100	490	590
1050	30" x 60"				
1060	Minimum	EA.	58.00	260	318
1062	Average	"	75.00	420	495
1064	Maximum	"	100	520	620
1160	Casement				
1180	1 leaf, 22" x 38" high				
1220	Minimum	EA.	52.00	290	342
1222	Average	"	65.00	350	415
1224	Maximum	"	87.00	410	497
1230	2 leaf, 50" x 50" high				
1240	Minimum	EA.	65.00	780	845
1242	Average	"	87.00	1,010	1,097
1244	Maximum	"	130	1,160	1,290
1250	3 leaf, 71" x 62" high				
1260	Minimum	EA.	65.00	1,280	1,345
1262	Average	"	87.00	1,300	1,387
1264	Maximum	"	130	1,560	1,690
1290	5 leaf, 119" x 75" high				

		UNIT	LABOR	MAT.	TOTAL
08600.10	**WOOD WINDOWS, Cont'd...**				
1300	Minimum	EA.	75.00	2,200	2,275
1302	Average	"	100	2,370	2,470
1304	Maximum	"	170	3,040	3,210
1360	Picture window, fixed glass, 54" x 54" high				
1400	Minimum	EA.	65.00	460	525
1422	Average	"	75.00	510	585
1424	Maximum	"	87.00	910	997
1430	68" x 55" high				
1440	Minimum	EA.	65.00	820	885
1442	Average	"	75.00	940	1,015
1444	Maximum	"	87.00	1,230	1,317
1480	Sliding, 40" x 31" high				
1520	Minimum	EA.	52.00	270	322
1522	Average	"	65.00	410	475
1524	Maximum	"	87.00	500	587
1530	52" x 39" high				
1540	Minimum	EA.	65.00	340	405
1542	Average	"	75.00	500	575
1544	Maximum	"	87.00	560	647
1550	64" x 72" high				
1560	Minimum	EA.	65.00	520	585
1562	Average	"	87.00	830	917
1564	Maximum	"	100	920	1,020
1760	Awning windows				
1780	34" x 21" high				
1800	Minimum	EA.	52.00	290	342
1822	Average	"	65.00	330	395
1824	Maximum	"	87.00	390	477
1880	48" x 27" high				
1900	Minimum	EA.	58.00	360	418
1902	Average	"	75.00	420	495
1904	Maximum	"	100	500	600
1920	60" x 36" high				
1940	Minimum	EA.	65.00	370	435
1942	Average	"	87.00	660	747
1944	Maximum	"	100	750	850
8000	Window frame, milled				
8010	Minimum	L.F.	10.50	5.28	15.78
8020	Average	"	13.00	5.88	18.88
8030	Maximum	"	17.50	8.85	26.35
08710.10	**HINGES**				
1200	Hinges				
1250	3 x 3 butts, steel, interior, plain bearing	PAIR			20.75
1260	4 x 4 butts, steel, standard	"			30.50
1270	5 x 4-1/2 butts, bronze/s. steel, heavy duty	"			79.00
1290	Pivot hinges				
1300	Top pivot	EA.			88.00
1310	Intermediate pivot	"			94.00
1320	Bottom pivot	"			180

		UNIT	LABOR	MAT.	TOTAL
08710.20	**LOCKSETS**				
1280	Latchset, heavy duty				
1300	Cylindrical	EA.	32.50	160	193
1320	Mortise	"	52.00	160	212
1325	Lockset, heavy duty				
1330	Cylindrical	EA.	32.50	250	283
1350	Mortise	"	52.00	280	332
2200	Preassembled locks and latches, brass				
2220	Latchset, passage or closet latch	EA.	43.50	250	294
2225	Lockset				
2230	Privacy (bath or bathroom)	EA.	43.50	300	344
2240	Entry lock	"	43.50	430	474
2285	Lockset				
2290	Privacy (bath or bedroom)	EA.	43.50	190	234
2300	Entry lock	"	43.50	220	264
08710.30	**CLOSERS**				
2600	Door closers				
2610	Standard	EA.	65.00	220	285
2620	Heavy duty	"	65.00	260	325
08710.40	**DOOR TRIM**				
1600	Panic device				
1610	Mortise	EA.	130	780	910
1620	Vertical rod	"	130	1,170	1,300
1630	Labeled, rim type	"	130	810	940
1640	Mortise	"	130	1,060	1,190
1650	Vertical rod	"	130	1,130	1,260
2300	Door plates				
2305	Kick plate, aluminum, 3 beveled edges				
2310	10" x 28"	EA.	26.00	22.00	48.00
2340	10" x 38"	"	26.00	28.50	54.50
2350	Push plate, 4" x 16"				
2360	Aluminum	EA.	10.50	22.00	32.50
2371	Bronze	"	10.50	84.00	94.50
2380	Stainless steel	"	10.50	67.00	77.50
2385	Armor plate, 40" x 34"	"	20.75	77.00	97.75
2388	Pull handle, 4" x 16"				
2390	Aluminum	EA.	10.50	94.00	105
2400	Bronze	"	10.50	180	191
2420	Stainless steel	"	10.50	140	151
2425	Hasp assembly				
2430	3"	EA.	8.70	4.40	13.10
2440	4-1/2"	"	11.50	5.50	17.00
2450	6"	"	15.00	8.74	23.74
08710.60	**WEATHERSTRIPPING**				
0100	Weatherstrip, head and jamb, metal strip, neoprene bulb				
0140	Standard duty	L.F.	2.90	4.95	7.85
0160	Heavy duty	"	3.26	5.50	8.76
3980	Spring type				
4000	Metal doors	EA.	130	55.00	185
4010	Wood doors	"	170	55.00	225
4020	Sponge type with adhesive backing	"	52.00	51.00	103
4500	Thresholds				

		UNIT	LABOR	MAT.	TOTAL
08710.60	**WEATHERSTRIPPING, Cont'd...**				
4510	Bronze	L.F.	13.00	53.00	66.00
4515	Aluminum				
4520	Plain	L.F.	13.00	36.50	49.50
4525	Vinyl insert	"	13.00	39.25	52.25
4530	Aluminum with grit	"	13.00	37.50	50.50
4533	Steel				
4535	Plain	L.F.	13.00	29.75	42.75
4540	Interlocking	"	43.50	39.50	83.00
08810.10	**GLAZING**				
0800	Sheet glass, 1/8" thick	S.F.	3.22	7.70	10.92
1020	Plate glass, bronze or grey, 1/4" thick	"	5.28	11.25	16.53
1040	Clear	"	5.28	8.80	14.08
1060	Polished	"	5.28	10.50	15.78
1980	Plexiglass				
2000	1/8" thick	S.F.	5.28	4.95	10.23
2020	1/4" thick	"	3.22	8.91	12.13
3000	Float glass, clear				
3010	3/16" thick	S.F.	4.84	5.99	10.83
3020	1/4" thick	"	5.28	6.10	11.38
3040	3/8" thick	"	7.26	12.25	19.51
3100	Tinted glass, polished plate, twin ground				
3120	3/16" thick	S.F.	4.84	8.25	13.09
3130	1/4" thick	"	5.28	8.25	13.53
3140	3/8" thick	"	7.26	13.25	20.51
5000	Insulated glass, bronze or gray				
5020	1/2" thick	S.F.	9.68	15.25	24.93
5040	1" thick	"	14.50	18.25	32.75
5100	Spandrel glass, polished bronze/grey, 1 side, 1/4" thick	"	5.28	12.25	17.53
6000	Tempered glass (safety)				
6010	Clear sheet glass				
6020	1/8" thick	S.F.	3.22	8.41	11.63
6030	3/16" thick	"	4.47	10.25	14.72
6040	Clear float glass				
6050	1/4" thick	S.F.	4.84	8.80	13.64
6060	5/16" thick	"	5.81	15.75	21.56
6070	3/8" thick	"	7.26	19.25	26.51
6080	1/2" thick	"	9.68	26.25	35.93
6800	Insulating glass, two lites, clear float glass				
6840	1/2" thick	S.F.	9.68	12.00	21.68
6850	5/8" thick	"	11.50	13.75	25.25
6860	3/4" thick	"	14.50	15.25	29.75
6870	7/8" thick	"	16.50	16.00	32.50
6880	1" thick	"	19.25	17.00	36.25
6885	Glass seal edge				
6890	3/8" thick	S.F.	9.68	10.00	19.68
6895	Tinted glass				
6900	1/2" thick	S.F.	9.68	20.50	30.18
6910	1" thick	"	19.25	22.00	41.25
6920	Tempered, clear				
6930	1" thick	S.F.	19.25	40.25	59.50
7200	Plate mirror glass				

		UNIT	LABOR	MAT.	TOTAL
08810.10	**GLAZING, Cont'd...**				
7205	1/4" thick				
7210	15 sf	S.F.	5.81	10.50	16.31
7220	Over 15 sf	"	5.28	9.68	14.96
9650	Sand-Blasted Glass				
9660	3/16" Float glass, full sandblast, no custom decoration	S.F.	4.84	10.50	15.34
9670	3/8" Float glass, full sandblast, no custom decoration	"	5.81	12.25	18.06
9680	Beveled glass				
9690	commercial standard grade	S.F.	5.81	130	136
08910.10	**GLAZED CURTAIN WALLS**				
1000	Curtain wall, aluminum system, framing sections				
1005	2" x 3"				
1010	Jamb	L.F.	4.84	11.00	15.84
1020	Horizontal	"	4.84	11.25	16.09
1030	Mullion	"	4.84	15.00	19.84
1035	2" x 4"				
1040	Jamb	L.F.	7.26	15.00	22.26
1060	Horizontal	"	7.26	15.50	22.76
1070	Mullion	"	7.26	15.00	22.26
1080	3" x 5-1/2"				
1090	Jamb	L.F.	7.26	19.75	27.01
1100	Horizontal	"	7.26	22.00	29.26
1110	Mullion	"	7.26	20.00	27.26
1115	4" corner mullion	"	9.68	26.50	36.18
1120	Coping sections				
1130	1/8" x 8"	L.F.	9.68	27.50	37.18
1140	1/8" x 9"	"	9.68	27.75	37.43
1150	1/8" x 12-1/2"	"	11.50	28.50	40.00
1160	Sill section				
1170	1/8" x 6"	L.F.	5.81	27.25	33.06
1180	1/8" x 7"	"	5.81	27.50	33.31
1190	1/8" x 8-1/2"	"	5.81	28.00	33.81
1200	Column covers, aluminum				
1210	1/8" x 26"	L.F.	14.50	27.25	41.75
1220	1/8" x 34"	"	15.25	27.50	42.75
1230	1/8" x 38"	"	15.25	27.75	43.00
1500	Doors				
1600	Aluminum framed, standard hardware				
1620	Narrow stile				
1630	2-6 x 7-0	EA.	290	620	910
1640	3-0 x 7-0	"	290	620	910
1660	3-6 x 7-0	"	290	640	930
1700	Wide stile				
1720	2-6 x 7-0	EA.	290	1,060	1,350
1730	3-0 x 7-0	"	290	1,140	1,430
1750	3-6 x 7-0	"	290	1,220	1,510
1800	Flush panel doors, to match adjacent wall panels				
1810	2-6 x 7-0	EA.	360	890	1,250
1820	3-0 x 7-0	"	360	940	1,300
1840	3-6 x 7-0	"	360	970	1,330
2100	Wall panel, insulated				
2120	"U"=.08	S.F.	4.84	11.25	16.09

		UNIT	LABOR	MAT.	TOTAL
08910.10	**GLAZED CURTAIN WALLS, Cont'd...**				
2140	"U"=.10	S.F.	4.84	10.50	15.34
2160	"U"=.15	"	4.84	9.57	14.41
3000	Window wall system, complete				
3010	Minimum	S.F.	5.81	29.75	35.56
3030	Average	"	6.45	47.50	53.95
3050	Maximum	"	8.30	110	118
4860	Added costs				
4870	For bronze, add 20% to material				
4880	For stainless steel, add 50% to material				

		UNIT	COST
08999.10	**DOORS**		
0900	HOLLOW METAL		
1000	CUSTOM FRAMES (16 ga)		
1010	2'6" x 6'8" x 4 3/4"	EA.	337
1020	2'8" x 6'8" x 4 3/4"	"	350
1030	3'0" x 6'8" x 4 3/4"	"	380
1040	3'4" x 6'8" x 4 3/4"	"	390
2000	Add for A, B or C Label	"	42.25
2010	Add for Side Lights	"	196
2020	Add for Frames over 7'0" - Height	"	78.25
2030	Add for Frames over 6 3/4" - Width	"	68.50
3000	CUSTOM DOORS (1 3/8" or 1 3/4") (18 ga)		
3010	2'6" x 6'8" x 1 3/4"	EA.	500
3020	2'8" x 6'8" x 1 3/4"	"	520
3030	3'0" x 6'8" x 1 3/4"	"	540
3040	3'4" x 6'8" x 1 3/4"	"	550
4000	Add for 16 ga	"	78.25
4010	Add for B and C Label	"	34.50
4020	Add for Vision Panels or Lights	"	122
5000	STOCK FRAMES (16 ga)		
5010	2'6" x 6'8" or 7'0" x 4 3/4"	EA.	224
5020	2'8" x 6'8" or 7'0" x 4 3/4"	"	234
5030	3'0" x 6'8" or 7'0" x 4 3/4"	"	244
5040	3'4" x 6'8" or 7'0" x 4 3/4"	"	264
5050	Add for A, B or C Label	"	41.00
6000	STOCK DOORS (1 3/8" or 1 3/4" - 18 ga)		
6010	2'6" x 6'8" or 7'0"	EA.	530
6020	2'8" x 6'8" or 7'0"	"	550
6030	3'0" x 6'8" or 7'0"	"	550
6040	3'4" x 6'8" or 7'0"	"	600
6050	Add for B or C Label	"	32.25
08999.20	**WOOD DOORS (Incl. butts and locksets)**		
1000	FLUSH DOORS - No Label - Paint Grade		
2000	Birch - 7 Ply 2'4" x 6'8" x 1 3/8" -Hollow Core	EA.	224
2010	2'6" x 6'8" x 1 3/8"	"	224
2020	2'8" x 6'8" x 1 3/8"	"	244
2030	3'0" x 6'8" x 1 3/8"	"	234
2040	2'4" x 6'8" x 1 3/4"	"	244
2050	2'6" x 6'8" x 1 3/4"	"	244
2060	2'8" x 6'8" x 1 3/4"	"	254
2070	3'0" x 6'8" x 1 3/4"	"	264
2080	3'4" x 6'8" x 1 3/4"	"	264
2090	3'6" x 6'8" x 1 3/4"	"	264
3000	Birch - 7 Ply 2'4" x 6'8" x 1 3/8" -Solid Core	"	264
3010	2'6" x 6'8" x 1 3/8"	"	264
3020	2'8" x 6'8" x 1 3/8"	"	304
3030	3'0" x 6'8" x 1 3/8"	"	304
3040	2'4" x 6'8" x 1 3/4"	"	304
3050	2'6" x 6'8" x 1 3/4"	"	304
3060	2'8" x 6'8" x 1 3/4"	"	329
3070	3'0" x 6'8" x 1 3/4"	"	329
3080	3'4" x 6'8" x 1 3/4"	"	339

		UNIT	COST
08999.20	**WOOD DOORS (Incl. butts and locksets), Cont'd...**		
3090	3'6" x 6'8" x 1 3/4"E	EA.	390
4000	Add for Stain Grade	"	43.25
4010	Add for 5 Ply	"	93.00
4020	Add for Jamb & Trim - Solid - Knock Down	"	96.00
4030	Add for Jamb & Trim - Veneer - Knock Down	"	63.50
4040	Add for Red Oak - Rotary Cut	"	38.75
4050	Add for Red Oak - Plain Sliced	"	52.00
5000	Deduct for Lauan	"	41.58
5010	Add for Architectural Grade - 7 Ply	"	90.25
5020	Add for Vinyl Overlay	"	69.75
5030	Add for 7' - 0" Doors	"	30.58
5040	Add for Lite Cutouts with Metal Frame	"	102
5050	Add for Wood Louvres	"	155
5060	Add for Transom Panels & Side Panels	"	196
6000	LABELED - 1 3/4" - Paint Grade		
6010	2'6" x 6'8" - Birch - 20 Min.	EA.	339
6020	2'8" x 6'8"	"	319
6030	3'0" x 6'8"	"	329
6040	3'4" x 6'8"	"	370
6050	3'6" x 6'8"	"	420
7000	2'6" x 6'8" - Birch - 45 Min	"	350
7010	2'8" x 6'8"	"	370
7020	3'0" x 6'8"	"	380
7030	3'4" x 6'8"	"	400
7040	3'6" x 6'8"	"	410
8000	2'6" x 6'8" - Birch - 60 Min	"	470
8010	2'8" x 6'8"	"	500
8020	3'0" x 6'8"	"	510
8030	3'4" x 6'8"	"	540
8040	3'6" x 6'8"	"	429
9000	PANEL DOORS		
9010	Exterior - 1 3/4" -Pine		
9020	2'8" x 6'8"	EA.	810
9030	3'0" x 6'8"	"	780
9040	Interior - 1 3/8" -Pine		
9050	2'6" x 6'8"	EA.	410
9060	2'8" x 6'8"	"	430
9070	3'0" x 6'8"	"	460
9080	Add for Sidelights	"	360
9140	Exterior - 1 3/4"- Birch/ Oak		
9150	2'8" x 6'8"	EA.	840
9160	3'0" x 6'8"	"	830
9170	Interior - 1 3/8"		
9180	2'6" x 6'8"	EA.	550
9190	2'8" x 6'8"	"	570
9200	3'0" x 6'8"	"	610
9210	Add for Sidelights	"	560
9212	LOUVERED DOORS - 1 3/8" -Pine		
9213	2'0" x 6'8"	EA.	300
9214	2'6" x 6'8"	"	350
9215	2'8" x 6'8"	"	370
9216	3'0" x 6'8"	"	390

		UNIT	COST
08999.20	**WOOD DOORS (incl. butts and locksets), Cont'd...**		
9220	Birch/Oak - 1 3/8"		
9230	2'0" x 6'8"	EA.	650
9240	2'6" x 6'8"	"	660
9250	2'8" x 6'8"	"	690
9260	3'0" x 6'8"	"	700
9270	BI-FOLD with hdwr. 1 3/8" - Flush		
9280	Birch/Oak, 2'0" x 6'8"	EA.	224
9290	2'4" x 6'8"	"	249
9300	2'6" x 6'8"	"	269
9320	Lauan, 2'0" x 6'8"	"	224
9330	2'4" x 6'8"	"	234
9340	2'6" x 6'8"	"	244
9350	Add for Prefinished	"	28.58
9360	CAFE DOORS - 1 1/8" - Pair		
9370	2'6" x 3'8"	EA.	460
9380	2'8" x 3'8"	"	470
9390	3'0" x 3'8"	"	490
9400	FRENCH DOORS - 1 3/8" -Pine		
9410	2'6" x 3'8"	EA.	600
9420	2'8" x 3'8"	"	610
9430	3'0" x 3'8"	"	640
9440	2'6" x 3'8" -Birch/Oak	"	790
9450	2'8" x 3'8"	"	800
9460	3'0" x 3'8"	"	810
9470	DUTCH DOORS		
9480	2'6" x 3'8"	EA.	900
9490	2'8" x 3'8"	"	1,000
9500	3'0" x 3'8"	"	1,260
9510	PREHUNG DOORS (INCL. TRIM)		
9520	Exterior, entrance - 1 3/4"		
9530	Panel - 2'8" x 6'8"	EA.	780
9540	3'0" x 6'8"	"	860
9550	3'0" x 7'0"	"	940
9560	Add for Insulation	"	264
9570	Add for Side Lights	"	324
9580	Interior - 1 3/8"		
9590	Flush, H.C. - 2'6" x 6'8" -Birch/Oak	EA.	420
9600	2'8" x 6'8"	"	430
9610	3'0" x 6'8"Each	"	450
9620	Flush, H.C. - 2'6" x 6'8" -Lauan	"	329
9630	2'8" x 6'8"	"	360
9640	3'0" x 6'8"Each	"	360
9650	Add for Int. Panel Door	"	299
08999.30	**SPECIAL DOORS**		
1000	BI-FOLDING - PREHUNG, 1 3/8"		
1020	Wood - 2 Door - 2'0" x 6'8" -Oak/flush	EA.	329
1030	2'6" x 6'8"	"	370
1040	3'0" x 6'8"	"	390
1050	4 Door - 4'0" x 6'8"	"	420
1060	5'0" x 6'8"	"	430
1070	6'0" x 6'8"	"	510

		UNIT	COST
08999.30	**SPECIAL DOORS, Cont'd...**		
1080	Add for Prefinished	EA.	39.75
1090	Add for Jambs and Casings	"	97.00
2000	Wood - 2 Door - 2'0" x 6'8" Pine/Panel	"	650
2010	2'6" x 6'8"	"	660
2020	3'0" x 6'8"	"	770
2030	4 Door - 4'0" x 6'8"	"	970
2040	5'0" x 6'8"	"	1,050
2050	6'0" x 6'8"	"	1,270
3000	Metal - 2 Door - 2'0" x 6'8"	"	206
3010	2'6" x 6'8"	"	206
3020	3'0" x 6'8"	"	254
3030	4 Door - 4'0" x 6'8"	"	274
3040	5'0" x 6'8"	"	299
3050	6'0" x 6'8"	"	319
3060	Add for Plastic Overlay	"	46.25
3070	Add for Louvre or Decorative Type	"	240
4000	Leaded Mirror (Based on 2-Panel Unit)		
4010	4'0" x 6'8"	EA.	940
4020	5'0" x 6'8"	"	1,090
4030	6'0" x 6'8"	"	1,090
7000	ROLLING - DOORS & GRILLES		
7010	Doors - 8' x 8'	EA.	3,770
7020	10' x 10'	"	3,950
7030	Grilles - 6'-8" x 3'-2"	"	1,310
7040	6'-8" x 4'-2"	"	1,480
8000	SHOWER DOORS - 28" x 66"	"	390
9000	SLIDING OR PATIO DOORS		
9010	Metal (Aluminum) - Including Glass Thresholds & Screen		
9020	Opening 8'0" x 6'8"	EA.	2,090
9030	8'0" x 8'0"	"	2,360
9040	Wood		
9050	Vinyl Clad 6'0" x 6'10"	EA.	2,600
9060	8'0" x 6'10"	"	2,860
9070	Pine - Prefinished 6'0" x 6'10"	"	2,580
9080	8'0" x 6'10"	"	2,870
9090	Add for Grilles	"	420
9100	Add for Triple Glazing	"	274
9110	Add for Screen	"	196
9120	SOUND REDUCTION - Metal	"	2,510
9130	Wood	"	1,650
9200	VAULT DOORS - 6'6" x 2'2" with Frame - 2 hour	"	4,290
08999.50	**METAL WINDOWS**		
1000	EACH, 2'4" x 4'6" - ALUMINUM WINDOWS		
1010	Casement and Awning	EA.	494
1020	Sliding or Horizontal	"	406
1030	Double and Single-Hung	"	436
1040	or Vertical Sliding		350
1050	Projected	EA.	464
1060	Add for Screens	"	66.70
1070	Add for Storms	"	117
1080	Add for Insulated Glass	"	102

		UNIT	COST
08999.50	**METAL WINDOWS, Cont'd...**		
2000	ALUMINUM SASH		
2010	Casement	EA.	426
2020	Sliding	"	376
2030	Single-Hung	"	376
2040	Projected	"	315
2050	Fixed	"	325
3000	STEEL WINDOWS		
3010	Double-Hung	EA.	829
3020	Projected	"	799
4000	STEEL SASH		
4010	Casement	EA.	614
4020	Double-Hung	"	674
4030	Projected	"	524
4040	Fixed	"	524
5000	By S.F., 2'4" x 4'6" - ALUMINUM WINDOWS		
5010	Casement and Awning	S.F.	46.12
5020	Sliding or Horizontal	"	36.20
5030	Double and Single-Hung	"	40.82
5040	or Vertical Sliding	"	33.50
5050	Projected	"	45.94
5060	Add for Screens	"	6.53
5070	Add for Storms	"	11.87
5080	Add for Insulated Glass	"	9.91
6000	ALUMINUM SASH		
6010	Casement	S.F.	43.12
6020	Sliding	"	36.45
6030	Single-Hung	"	36.45
6040	Projected	"	30.95
6050	Fixed	"	32.12
7000	STEEL WINDOWS		
7010	Double-Hung	S.F.	67.99
7020	Projected	"	64.58
8000	STEEL SASH		
8010	Casement	S.F.	52.47
8020	Double-Hung	"	64.58
8030	Projected	"	44.94
8040	Fixed	"	46.41
08999.60	**WOOD WINDOWS**		
1000	BASEMENT OR UTILITY		
1020	Prefinished with Screen, 2'8" x 1'4"	EA.	255
1030	2'8" x 2'0"	"	296
1040	Add for Double Glazing or Storm	"	65.50
2000	CASEMENT OR AWNING		
2010	Operating Units - Insulating Glass		
2020	Single 2'4" x 4'0"	EA.	720
2030	2'4" x 5'0"	"	680
3000	Double with Screen 4'0" x 4'0" (2 units)	"	920
3010	4'0" x 5'0" (2 units)	"	1,170
3020	Triple with Screen 6'0" x 4'0" (3 units)	"	1,600
3030	6'0" x 5'0" (3 units)	"	1,610
4000	Fixed Units - Insulating Glass		

		UNIT	COST
08999.60	**WOOD WINDOWS, Cont'd...**		
4010	Single 2'4" x 4'0"	EA.	570
4020	2'4" x 5'0"	"	640
4030	Picture 4'0" x 4'6"	"	790
4040	4'0" x 6'0"	"	940
5000	DOUBLE HUNG - Insul. Glass - 2'6" x 3'6"	"	610
5010	2'6" x 4'2"	"	610
5020	3'2" x 3'6"	"	630
5030	3'2" x 4'2"	"	700
5040	Add for Triple Glazing	"	98.50
6000	GLIDER - & Insul. Glass with Screen - 4'0" x 3'6"	"	1,420
6010	5'0" x 4'0"	"	1,600
8000	PICTURE WINDOWS		
8010	9'6" x 4'10"	EA.	2,990
8020	9'6" x 5'6"	"	2,280
9000	CASEMENT - Insulating Glass		
9010	30° - 5'10" x 4'2" - (3 units)	EA.	2,170
9015	7'10" x 4'2" - (4 units)	"	2,630
9020	5'10" x 5'2" - (3 units)	"	2,270
9030	7'10" x 5'2" - (4 units)	"	2,580
9040	45° - 5'4" x 4'2" - (3 units)	"	2,160
9050	7'4" x 4'2" - (4 units)	"	2,380
9060	5'4" x 5'2" - (3 units)	"	2,520
9070	7'4" x 5'2" - (4 units)	"	2,840
9100	CASEMENT BOW WINDOWS - Insulating Glass		
9110	6'2" x 4'2" - (3 units)	EA.	2,180
9120	8'2" x 4'2" - (4 units)	"	2,880
9130	6'2" x 5'2" - (3 units)	"	2,430
9140	8'2" x 5'2" - (4 units)	"	3,170
9150	Add for Triple Glazing - per unit	"	81.50
9160	Add for Bronze Glazing - per unit	"	81.50
9170	Deduct for Primed Only - per unit	"	20.44
9180	Add for Screens - per unit	"	25.10
9200	90° BOX BAY WINDOWS - Insulating Glass 4'8" x 4'2"	"	2,880
9210	6'8" x 4'2"	"	3,660
9220	6'8" x 5'2"	"	3,800
9300	ROOF WINDOWS 1'10" x 3'10"-Fixed	"	840
9310	2'4" x 3'10"	"	960
9320	3'8" x 3'10"	"	1,180
9400	ROOF WINDOWS 1'10" x 3'10"-Movable	"	1,120
9410	2'4" x 3'10"	"	1,380
9420	3'8" x 3'10"	"	1,590
9500	CIRCULAR TOPS & ROUNDS - 4'0"	"	900
9510	6'0"	"	2,000
08999.70	**SPECIAL WINDOWS**		
1000	LIGHT-PROOF WINDOWS	S.F.	54.25
2000	PASS WINDOWS	"	40.96
3000	DETENTION WINDOWS	"	56.75
4000	VENETIAN BLIND WINDOWS (ALUMINUM)	"	47.42
5000	SOUND-CONTROL WINDOWS	"	46.24

		UNIT	COST
08999.80	**DOOR AND WINDOW ACCESSORIES**		
1000	STORMS AND SCREENS		
1010	Windows		
1020	Screen Only - Wood - 3' x 5'	EA.	148
1030	Aluminum - 3' x 5'	"	162
1040	Storm & Screen Combination - Aluminum	"	184
1050	Doors	"	
1060	Screen Only - Wood	"	344
1070	Aluminum - 3' x 6' - 8'	"	344
1080	Storm & Screen Combination - Aluminum	"	404
1090	Wood 1 1/8"	"	459
2000	DETENTION SCREENS		
2010	Example: 4' 0" x 7' 0"	EA.	1,020
3000	DOOR OPENING ASSEMBLIES		
3010	Floor or Overhead Electric Eye Units		
3020	Swing - Single 3' x 7' Door - Hydraulic	EA.	5,660
3030	Double 6' x 7' Door - Hydraulic	"	8,970
3040	Sliding - Single 3' x 7' Door - Hydraulic	"	6,930
3050	Double 5' x 7' Doors- Hydraulic	"	8,930
3060	Industrial Doors - 10' x 8'	"	10,050
4000	SHUTTERS		
4010	16" x 1 1/8" x 48"	EA.	160
4020	16" x 1 1/8" x 60"	"	184
4030	16" x 1 1/8" x 72"	"	214
08999.90	**FINISH HARDWARE**		
1000	BUTT HINGES -Painted		
1010	3" x 3"	EA.	35.25
1020	3 1/2" x 3 1/2"	"	37.00
1030	4" x 4"	"	39.50
1040	4 1/2" x 4 1/2"	"	43.00
1050	4" x 4" Ball Bearing	"	65.00
1060	4 1/2" x 4 1/2" Ball Bearing	"	69.00
2000	Bronze		
2010	3" x 3"	EA.	37.00
2020	3 1/2" x 3 1/2"	"	37.75
2030	4" x 4"	"	39.50
2040	4 1/2" x 4 1/2"	"	45.75
2050	4" x 4" Ball Bearing	"	67.00
2060	4 1/2" x 4 1/2" Ball Bearing	"	76.00
3000	Chrome		
3010	3" x 3"	EA.	39.50
3020	3 1/2" x 3 1/2"	"	42.25
3030	4" x 4"	"	44.00
3040	4 1/2" x 4 1/2"	"	53.00
3050	4" x 4" Ball Bearing	"	76.00
3060	4 1/2" x 4 1/2" Ball Bearing	"	80.00
5000	CLOSERS - SURFACE MOUNTED - 3' - 0" Door	"	259
5010	3' - 4"	"	269
5020	3' - 8"	"	279
5030	4' - 0"	"	350
6000	CLOSERS - CONCEALED - Interior	"	430
6010	ExteriorEach	"	570

		UNIT	COST
08999.90	**FINISH HARDWARE, Cont'd...**		
6020	Add for Fusible Link - Electric	EA.	249
6030	CLOSERS - FLOOR HINGES - Interior	"	640
6040	Exterior	"	980
6050	Add for Hold Open Feature	"	102
6060	Add for Double Acting Feature	"	330
7000	DEAD BOLT LOCK - Cylinder - Outside Key	"	219
7010	Cylinder - Double Key	"	310
7020	Flush - Push/ Pull	"	85.00
8000	EXIT DEVICES (PANIC) - Surface	"	930
8010	Mortise Lock	"	1,190
8020	Concealed	"	2,960
8030	Handicap (ADA) Automatic	"	2,170
9000	HINGES, SPRING (PAINTED) - 6" Single Acting	"	160
9010	6" Double Acting	"	194
9101	LATCHSETS -Bronze or Chrome	"	259
9102	Stainless Steel	"	290
9200	LOCKSETS - Mortise - Bronze or Chrome - H.D.	"	350
9210	Mortise - Bronze or Chrome - S.D.	"	279
9220	Stainless Steel	"	460
9230	Cylindrical - Bronze or Chrome	"	310
9240	Stainless Steel	"	360
9300	LEVER HANDICAP - Latch Set	"	360
9310	Lock Set	"	430
9400	PLATES - Kick - 8" x 34" - Aluminum	"	112
9410	Bronze	"	103
9420	Push - 6" x 15" - Aluminum	"	74.25
9430	Bronze	"	105
9440	Push & Pull Combination - Aluminum	"	123
9450	Bronze	"	176
9500	STOPS AND HOLDERS		
9510	Holder - Magnetic (No Electric)	EA.	249
9520	Bumper	"	58.50
9530	Overhead - Bronze, Chrome or Aluminum	"	137
9540	Wall Stops	"	51.75
9550	Floor Stops	"	64.50
08999.91	**WEATHERSTRIPPING**		
1000	ASTRAGALS - Aluminum - 1/8" x 2"	EA.	78.25
1010	Painted Steel	"	70.25
2000	DOORS (WOOD) - Interlocking	"	112
2010	Spring Bronze	"	122
2020	Add for Metal Doors	"	56.00
3000	SWEEPS - 36" WOOD DOORS - Aluminum	"	31.75
3010	Vinyl	"	31.50
4000	THRESHOLDS - Aluminum - 4" x 1/2"	"	67.25
4010	Bronze 4" x 1/2"	"	84.75
4020	5 1/2" x 1/2"	"	89.75
5000	WINDOWS (WOOD) - Interlocking	"	95.25
5010	Spring Bronze	"	113

TABLE OF CONTENTS   PAGE

09110.10	METAL STUDS	UNIT	LABOR	MAT.	TOTAL
0060	Studs, non load bearing, galvanized				
0130	3-5/8", 20 ga.				
0140	12" o.c.	S.F.	1.30	0.85	2.15
0142	16" o.c.	"	1.04	0.66	1.70
0144	24" o.c.	"	0.87	0.49	1.36
0170	25 ga.				
0180	12" o.c.	S.F.	1.30	0.57	1.87
0182	16" o.c.	"	1.04	0.46	1.50
0184	24" o.c.	"	0.87	0.35	1.22
0210	6", 20 ga.				
0220	12" o.c.	S.F.	1.63	1.19	2.82
0222	16" o.c.	"	1.30	0.88	2.18
0224	24" o.c.	"	1.08	0.71	1.79
0230	25 ga.				
0240	12" o.c.	S.F.	1.63	0.78	2.41
0242	16" o.c.	"	1.30	0.61	1.91
0244	24" o.c.	"	1.08	0.46	1.54
0980	Load bearing studs, galvanized				
0990	3-5/8", 16 ga.				
1000	12" o.c.	S.F.	1.30	1.55	2.85
1020	16" o.c.	"	1.04	1.43	2.47
1110	18 ga.				
1130	12" o.c.	S.F.	0.87	1.11	1.98
1140	16" o.c.	"	1.04	1.21	2.25
1980	6", 16 ga.				
2000	12" o.c.	S.F.	1.63	2.07	3.70
2001	16" o.c.	"	1.30	1.87	3.17
3000	Furring				
3160	On beams and columns				
3170	7/8" channel	L.F.	3.48	0.55	4.03
3180	1-1/2" channel	"	4.01	0.66	4.67
4460	On ceilings				
4470	3/4" furring channels				
4480	12" o.c.	S.F.	2.17	0.39	2.56
4490	16" o.c.	"	2.08	0.30	2.38
4495	24" o.c.	"	1.86	0.22	2.08
4500	1-1/2" furring channels				
4520	12" o.c.	S.F.	2.37	0.66	3.03
4540	16" o.c.	"	2.17	0.49	2.66
4560	24" o.c.	"	2.00	0.34	2.34
5000	On walls				
5020	3/4" furring channels				
5050	12" o.c.	S.F.	1.74	0.39	2.13
5100	16" o.c.	"	1.63	0.30	1.93
5150	24" o.c.	"	1.53	0.22	1.75
5200	1-1/2" furring channels				
5210	12" o.c.	S.F.	1.86	0.66	2.52
5220	16" o.c.	"	1.74	0.49	2.23
5230	24" o.c.	"	1.63	0.34	1.97

		UNIT	LABOR	MAT.	TOTAL
09205.10	**GYPSUM LATH**				
1070	Gypsum lath, 1/2" thick				
1090	Clipped	S.Y.	2.90	7.20	10.10
1110	Nailed	"	3.26	7.20	10.46
09205.20	**METAL LATH**				
0960	Diamond expanded, galvanized				
0980	2.5 lb., on walls				
1010	Nailed	S.Y.	6.52	4.22	10.74
1030	Wired	"	7.45	4.22	11.67
1040	On ceilings				
1050	Nailed	S.Y.	7.45	4.22	11.67
1070	Wired	"	8.70	4.22	12.92
2230	Stucco lath				
2240	1.8 lb.	S.Y.	6.52	4.96	11.48
2300	3.6 lb.	"	6.52	5.56	12.08
2310	Paper backed				
2320	Minimum	S.Y.	5.22	3.85	9.07
2400	Maximum	"	7.45	6.21	13.66
09205.60	**PLASTER ACCESSORIES**				
0120	Expansion joint, 3/4", 26 ga., galvanized, one piece	L.F.	1.30	1.48	2.78
2000	Plaster corner beads, 3/4", galvanized	"	1.49	0.41	1.90
2020	Casing bead, expanded flange, galvanized	"	1.30	0.56	1.86
2100	Expanded wing, 1-1/4" wide, galvanized	"	1.30	0.66	1.96
2500	Joint clips for lath	EA.	0.26	0.17	0.43
2580	Metal base, galvanized, 2-1/2" high	L.F.	1.74	0.75	2.49
2600	Stud clips for gypsum lath	EA.	0.26	0.17	0.43
2700	Tie wire galvanized, 18 ga., 25 lb. hank	"			47.00
8000	Sound deadening board, 1/4", nailed or clipped	S.F.	0.87	0.31	1.18
09210.10	**PLASTER**				
0980	Gypsum plaster, trowel finish, 2 coats				
1000	Ceilings	S.Y.	16.00	6.20	22.20
1020	Walls	"	15.00	6.20	21.20
1030	3 coats				
1040	Ceilings	S.Y.	22.25	8.60	30.85
1060	Walls	"	19.75	8.60	28.35
7000	On columns, add to installation, 50%	"			
7020	Chases, fascia, and soffits, add to installation, 50%	"			
7040	Beams, add to installation, 50%	"			
09220.10	**PORTLAND CEMENT PLASTER**				
2980	Stucco, portland, gray, 3 coat, 1" thick				
3000	Sand finish	S.Y.	22.25	7.75	30.00
3020	Trowel finish	"	23.25	7.75	31.00
3030	White cement				
3040	Sand finish	S.Y.	23.25	8.85	32.10
3060	Trowel finish	"	25.75	8.85	34.60
3980	Scratch coat				
4000	For ceramic tile	S.Y.	5.14	2.81	7.95
4020	For quarry tile	"	5.14	2.81	7.95
5000	Portland cement plaster				
5020	2 coats, 1/2"	S.Y.	10.25	5.58	15.83
5040	3 coats, 7/8"	"	12.75	6.66	19.41

09250.10	GYPSUM BOARD	UNIT	LABOR	MAT.	TOTAL
0220	1/2", clipped to				
0240	Metal furred ceiling	S.F.	0.58	0.36	0.94
0260	Columns and beams	"	1.30	0.33	1.63
0270	Walls	"	0.52	0.33	0.85
0280	Nailed or screwed to				
0290	Wood framed ceiling	S.F.	0.52	0.33	0.85
0300	Columns and beams	"	1.16	0.33	1.49
0400	Walls	"	0.47	0.33	0.80
1000	5/8", clipped to				
1020	Metal furred ceiling	S.F.	0.65	0.36	1.01
1040	Columns and beams	"	1.45	0.36	1.81
1060	Walls	"	0.58	0.36	0.94
1070	Nailed or screwed to				
1080	Wood framed ceiling	S.F.	0.65	0.36	1.01
1100	Columns and beams	"	1.45	0.36	1.81
1120	Walls	"	0.58	0.36	0.94
1122	Vinyl faced, clipped to metal studs				
1124	1/2"	S.F.	0.65	0.88	1.53
1126	5/8"	"	0.65	0.93	1.58
1130	Add for				
1140	Fire resistant	S.F.			0.11
1180	Water resistant	"			0.17
1200	Water and fire resistant	"			0.22
1220	Taping and finishing joints				
1222	Minimum	S.F.	0.34	0.04	0.38
1224	Average	"	0.43	0.06	0.49
1226	Maximum	"	0.52	0.09	0.61
5020	Casing bead				
5022	Minimum	L.F.	1.49	0.15	1.64
5024	Average	"	1.74	0.16	1.90
5026	Maximum	"	2.61	0.20	2.81
5040	Corner bead				
5042	Minimum	L.F.	1.49	0.16	1.65
5044	Average	"	1.74	0.20	1.94
5046	Maximum	"	2.61	0.25	2.86
09310.10	**CERAMIC TILE**				
0980	Glazed wall tile, 4-1/4" x 4-1/4"				
1000	Minimum	S.F.	3.60	2.32	5.92
1020	Average	"	4.21	3.68	7.89
1040	Maximum	"	5.05	13.25	18.30
1042	6" x 6"				
1044	Minimum	S.F.	3.15	1.41	4.56
1046	Average	"	3.60	1.90	5.50
1048	Maximum	"	4.21	2.37	6.58
2960	Base, 4-1/4" high				
2980	Minimum	L.F.	6.31	3.83	10.14
3000	Average	"	6.31	4.46	10.77
3040	Maximum	"	6.31	5.89	12.20
3042	Glazed modlings and trim, 12" x 12"				
3044	Minimum	L.F.	5.05	2.11	7.16
3046	Average	"	5.05	3.22	8.27

		UNIT	LABOR	MAT.	TOTAL
09310.10	**CERAMIC TILE, Cont'd...**				
3048	Maximum	L.F.	5.05	4.33	9.38
6100	Unglazed floor tile				
6120	Portland cement bed, cushion edge, face mounted				
6140	1" x 1"	S.F.	4.59	6.71	11.30
6150	2" x 2"	"	4.21	7.09	11.30
6162	4" x 4"	"	4.21	6.60	10.81
6164	6" x 6"	"	3.60	2.36	5.96
6166	12" x 12"	"	3.15	2.07	5.22
6168	16" x 16"	"	2.80	1.80	4.60
6170	18" x 18"	"	2.52	1.74	4.26
6200	Adhesive bed, with white grout				
6220	1" x 1"	S.F.	4.59	6.71	11.30
6230	2" x 2"	"	4.21	7.09	11.30
6260	4" x 4"	"	4.21	7.09	11.30
6262	6" x 6"	"	3.60	2.36	5.96
6264	12" x 12"	"	3.15	2.07	5.22
6266	16" x 16"	"	2.80	1.80	4.60
6268	18" x 18"	"	2.52	1.74	4.26
6300	Organic adhesive bed, thin set, back mounted				
6320	1" x 1"	S.F.	4.59	6.71	11.30
6350	2" x 2"	"	4.21	7.09	11.30
6360	For group 2 colors, add to material, 10%				
6370	For group 3 colors, add to material, 20%				
6380	For abrasive surface, add to material, 25%				
8990	Ceramic accessories				
9000	Towel bar, 24" long				
9004	Average	EA.	25.25	16.25	41.50
9020	Soap dish				
9024	Average	EA.	42.00	10.50	52.50
09330.10	**QUARRY TILE**				
1060	Floor				
1080	4 x 4 x 1/2"	S.F.	6.73	5.69	12.42
1120	6 x 6 x 3/4"	"	6.31	6.94	13.25
1200	Wall, applied to 3/4" portland cement bed				
1220	4 x 4 x 1/2"	S.F.	10.00	4.20	14.20
1240	6 x 6 x 3/4"	"	8.42	4.69	13.11
1320	Cove base				
1330	5 x 6 x 1/2" straight top	L.F.	8.42	5.50	13.92
1340	6 x 6 x 3/4" round top	"	8.42	5.11	13.53
1360	Stair treads 6 x 6 x 3/4"	"	12.75	7.53	20.28
1380	Window sill 6 x 8 x 3/4"	"	10.00	6.87	16.87
1400	For abrasive surface, add to material, 25%				
09410.10	**TERRAZZO**				
1100	Floors on concrete, 1-3/4" thick, 5/8" topping				
1120	Gray cement	S.F.	7.34	4.23	11.57
1140	White cement	"	7.34	4.62	11.96
1200	Sand cushion, 3" thick, 5/8" top, 1/4"				
1220	Gray cement	S.F.	8.56	5.00	13.56
1240	White cement	"	8.56	5.55	14.11
1260	Monolithic terrazzo, 3-1/2" base slab, 5/8" topping	"	6.42	3.97	10.39
1280	Terrazzo wainscot, cast-in-place, 1/2" thick	"	12.75	7.48	20.23

		UNIT	LABOR	MAT.	TOTAL
09410.10	**TERRAZZO, Cont'd...**				
1300	Base, cast in place, terrazzo cove type, 6" high	L.F.	7.34	8.86	16.20
1320	Curb, cast in place, 6" wide x 6" high, polished top	"	25.75	9.84	35.59
1400	For venetian type terrazzo, add to material, 10%				
1420	For abrasive heavy duty terrazzo, add to material, 15%				
1480	Divider strips				
1500	Zinc	L.F.			1.50
1510	Brass	"			2.80
09510.10	**CEILINGS AND WALLS**				
1520	Acoustical panels, suspension system not included				
1540	Fiberglass panels				
1550	5/8" thick				
1560	2' x 2'	S.F.	0.74	1.52	2.26
1580	2' x 4'	"	0.58	1.44	2.02
1590	3/4" thick				
1600	2' x 2'	S.F.	0.74	2.71	3.45
1620	2' x 4'	"	0.58	2.40	2.98
1640	Glass cloth faced fiberglass panels				
1660	3/4" thick	S.F.	0.87	2.50	3.37
1680	1" thick	"	0.87	2.79	3.66
1700	Mineral fiber panels				
1710	5/8" thick				
1720	2' x 2'	S.F.	0.74	1.12	1.86
1740	2' x 4'	"	0.58	1.12	1.70
1750	3/4" thick				
1760	2' x 2'	S.F.	0.74	1.51	2.25
1780	2' x 4'	"	0.58	1.51	2.09
3000	Acoustical tiles, suspension system not included				
3020	Fiberglass tile, 12" x 12"				
3040	5/8" thick	S.F.	0.94	1.39	2.33
3060	3/4" thick	"	1.16	1.61	2.77
3080	Glass cloth faced fiberglass tile				
3100	3/4" thick	S.F.	1.16	3.24	4.40
3120	3" thick	"	1.30	3.63	4.93
3140	Mineral fiber tile, 12" x 12"				
3150	5/8" thick				
3160	Standard	S.F.	1.04	0.82	1.86
3180	Vinyl faced	"	1.04	1.65	2.69
3190	3/4" thick				
3195	Standard	S.F.	1.04	1.21	2.25
3200	Vinyl faced	"	1.04	2.11	3.15
5500	Ceiling suspension systems				
5505	T bar system				
5510	2' x 4'	S.F.	0.52	1.15	1.67
5520	2' x 2'	"	0.58	1.25	1.83
5530	Concealed Z bar suspension system, 12" module	"	0.87	1.07	1.94
5550	For 1-1/2" carrier channels, 4' o.c., add	"			0.38
5560	Carrier channel for recessed light fixtures	"			0.69
09550.10	**WOOD FLOORING**				
0100	Wood strip flooring, unfinished				
1000	Fir floor				
1010	C and better				

			UNIT	LABOR	MAT.	TOTAL
09550.10	**WOOD FLOORING, Cont'd...**					
1020	Vertical grain		S.F.	1.74	3.52	5.26
1040	Flat grain		"	1.74	3.32	5.06
1060	Oak floor					
1080	Minimum		S.F.	2.48	3.71	6.19
1100	Average		"	2.48	5.12	7.60
1120	Maximum		"	2.48	7.42	9.90
1340	Added costs					
1350	For factory finish, add to material, 10%					
1355	For random width floor, add to total, 20%					
1360	For simulated pegs, add to total, 10%					
3000	Gym floor, 2 ply felt, 25/32" maple, finished, in mastic		S.F.	2.90	7.75	10.65
3020	Over wood sleepers		"	3.26	8.69	11.95
9020	Finishing, sand, fill, finish, and wax		"	1.30	0.66	1.96
9100	Refinish sand, seal, and 2 coats of polyurethane		"	1.74	1.15	2.89
9540	Clean and wax floors		"	0.26	0.19	0.45
09550.20	**BAMBOO FLOORING**					
0010	Vertical, Carbonized Medium, 3' vertical grain		S.F.	1.74	6.03	7.77
0020	Natural		"	1.74	5.94	7.68
0030	3' horizontal grain		"	1.74	6.03	7.77
0040	Natural		"	1.74	5.94	7.68
0050	3' Stained		"	1.74	5.81	7.55
0060	6' spice		"	1.74	5.81	7.55
0070	3' stained, butterscotch		"	1.74	5.81	7.55
0080	6' tiger		"	1.74	5.81	7.55
0090	3' stained, Irish moss		"	1.74	5.81	7.55
0100	Vice-lock, 12 mm., laminate flooring, maple		"	1.74	3.85	5.59
0101	Oak		"	1.74	3.30	5.04
0102	Pine		"	1.74	4.40	6.14
0103	Espresso		"	1.74	4.40	6.14
0104	Standard, hard maple		"	1.74	3.57	5.31
0105	Cherry		"	1.74	3.30	5.04
0106	Oak		"	1.74	2.20	3.94
0107	Walnut		"	1.74	2.20	3.94
0108	South pacific vice-lock, 12 mm, brazilian cherry		"	1.74	4.95	6.69
0109	Maple		"	1.74	4.40	6.14
0110	Teak		"	1.74	4.67	6.41
09630.10	**UNIT MASONRY FLOORING**					
1000	Clay brick					
1020	9 x 4-1/2 x 3" thick					
1040	Glazed		S.F.	4.35	8.03	12.38
1060	Unglazed		"	4.35	7.70	12.05
1070	8 x 4 x 3/4" thick					
1080	Glazed		S.F.	4.53	7.26	11.79
1100	Unglazed		"	4.53	7.15	11.68
1140	For herringbone pattern, add to labor, 15%					
09660.10	**RESILIENT TILE FLOORING**					
1020	Solid vinyl tile, 1/8" thick, 12" x 12"					
1040	Marble patterns		S.F.	1.30	4.45	5.75
1060	Solid colors		"	1.30	5.77	7.07
1080	Travertine patterns		"	1.30	6.49	7.79
2000	Conductive resilient flooring, vinyl tile					

		UNIT	LABOR	MAT.	TOTAL
09660.10	**RESILIENT TILE FLOORING, Cont'd...**				
2040	1/8" thick, 12" x 12"	S.F.	1.49	6.71	8.20
09665.10	**RESILIENT SHEET FLOORING**				
0980	Vinyl sheet flooring				
1000	Minimum	S.F.	0.52	3.83	4.35
1002	Average	"	0.63	6.19	6.82
1004	Maximum	"	0.87	10.50	11.37
1020	Cove, to 6"	L.F.	1.04	2.28	3.32
2000	Fluid applied resilient flooring				
2020	Polyurethane, poured in place, 3/8" thick	S.F.	4.35	10.50	14.85
6200	Vinyl sheet goods, backed				
6220	0.070" thick	S.F.	0.65	3.90	4.55
6260	0.125" thick	"	0.65	6.98	7.63
6280	0.250" thick	"	0.65	8.03	8.68
09678.10	**RESILIENT BASE AND ACCESSORIES**				
1000	Wall base, vinyl				
1130	4" high	L.F.	1.74	1.01	2.75
1140	6" high	"	1.74	1.37	3.11
09682.10	**CARPET PADDING**				
1000	Carpet padding				
1005	Foam rubber, waffle type, 0.3" thick	S.Y.	2.61	6.16	8.77
1010	Jute padding				
1022	Average	S.Y.	2.61	5.44	8.05
1030	Sponge rubber cushion				
1042	Average	S.Y.	2.61	6.60	9.21
1050	Urethane cushion, 3/8" thick				
1062	Average	S.Y.	2.61	5.77	8.38
09685.10	**CARPET**				
0990	Carpet, acrylic				
1000	24 oz., light traffic	S.Y.	5.80	14.50	20.30
1020	28 oz., medium traffic	"	5.80	17.50	23.30
2010	Nylon				
2020	15 oz., light traffic	S.Y.	5.80	20.25	26.05
2040	28 oz., medium traffic	"	5.80	26.50	32.30
2110	Nylon				
2120	28 oz., medium traffic	S.Y.	5.80	25.25	31.05
2140	35 oz., heavy traffic	"	5.80	30.75	36.55
2145	Wool				
2150	30 oz., medium traffic	S.Y.	5.80	41.75	47.55
2160	36 oz., medium traffic	"	5.80	44.00	49.80
2180	42 oz., heavy traffic	"	5.80	58.00	63.80
3000	Carpet tile				
3020	Foam backed				
3022	Minimum	S.F.	1.04	3.21	4.25
3024	Average	"	1.16	3.71	4.87
3026	Maximum	"	1.30	5.88	7.18
8980	Clean and vacuum carpet				
9000	Minimum	S.Y.	0.20	0.29	0.49
9020	Average	"	0.34	0.46	0.80
9040	Maximum	"	0.52	0.63	1.15

09905.10	PAINTING PREPARATION	UNIT	LABOR	MAT.	TOTAL
1000	Dropcloths				
1050	Minimum	S.F.	0.03	0.02	0.05
1100	Average	"	0.04	0.03	0.07
1150	Maximum	"	0.05	0.04	0.09
1200	Masking				
1250	Paper and tape				
1300	Minimum	L.F.	0.52	0.02	0.54
1350	Average	"	0.65	0.03	0.68
1400	Maximum	"	0.87	0.04	0.91
1450	Doors				
1500	Minimum	EA.	6.53	0.04	6.57
1550	Average	"	8.70	0.05	8.75
1600	Maximum	"	11.50	0.06	11.56
1650	Windows				
1700	Minimum	EA.	6.53	0.04	6.57
1750	Average	"	8.70	0.05	8.75
1800	Maximum	"	11.50	0.06	11.56
2000	Sanding				
2050	Walls and flat surfaces				
2100	Minimum	S.F.	0.34		0.34
2150	Average	"	0.43		0.43
2200	Maximum	"	0.52		0.52
2250	Doors and windows				
2300	Minimum	EA.	8.70		8.70
2350	Average	"	13.00		13.00
2400	Maximum	"	17.50		17.50
2450	Trim				
2500	Minimum	L.F.	0.65		0.65
2550	Average	"	0.87		0.87
2600	Maximum	"	1.16		1.16
2650	Puttying				
2700	Minimum	S.F.	0.80	0.01	0.81
2750	Average	"	1.04	0.02	1.06
2800	Maximum	"	1.30	0.03	1.33
09910.05	EXT. PAINTING, SITEWORK				
3000	Concrete Block				
3020	Roller				
3040	First Coat				
3060	Minimum	S.F.	0.26	0.16	0.42
3080	Average	"	0.34	0.16	0.50
3100	Maximum	"	0.52	0.16	0.68
3120	Second Coat				
3140	Minimum	S.F.	0.21	0.16	0.37
3160	Average	"	0.29	0.16	0.45
3180	Maximum	"	0.43	0.16	0.59
3200	Spray				
3220	First Coat				
3240	Minimum	S.F.	0.14	0.13	0.27
3260	Average	"	0.17	0.13	0.30
3280	Maximum	"	0.20	0.13	0.33
3300	Second Coat				

09910.05	EXT. PAINTING, SITEWORK, Cont'd...	UNIT	LABOR	MAT.	TOTAL
3320	Minimum	S.F.	0.09	0.13	0.22
3340	Average	"	0.11	0.13	0.24
3360	Maximum	"	0.16	0.13	0.29
3500	Fences, Chain Link				
3700	Roller				
3720	First Coat				
3740	Minimum	S.F.	0.37	0.11	0.48
3760	Average	"	0.43	0.11	0.54
3780	Maximum	"	0.49	0.11	0.60
3800	Second Coat				
3820	Minimum	S.F.	0.21	0.11	0.32
3840	Average	"	0.26	0.11	0.37
3860	Maximum	"	0.32	0.11	0.43
3880	Spray				
3900	First Coat				
3920	Minimum	S.F.	0.16	0.08	0.24
3940	Average	"	0.18	0.08	0.26
3960	Maximum	"	0.21	0.08	0.29
3980	Second Coat				
4000	Minimum	S.F.	0.12	0.08	0.20
4060	Average	"	0.14	0.08	0.22
4080	Maximum	"	0.16	0.08	0.24
4200	Fences, Wood or Masonry				
4220	Brush				
4240	First Coat				
4260	Minimum	S.F.	0.54	0.16	0.70
4280	Average	"	0.65	0.16	0.81
4300	Maximum	"	0.87	0.16	1.03
4320	Second Coat				
4340	Minimum	S.F.	0.32	0.16	0.48
4360	Average	"	0.40	0.16	0.56
4380	Maximum	"	0.52	0.16	0.68
4400	Roller				
4420	First Coat				
4440	Minimum	S.F.	0.29	0.16	0.45
4460	Average	"	0.34	0.16	0.50
4480	Maximum	"	0.40	0.16	0.56
4500	Second Coat				
4520	Minimum	S.F.	0.20	0.16	0.36
4540	Average	"	0.24	0.16	0.40
4560	Maximum	"	0.32	0.16	0.48
4580	Spray				
4600	First Coat				
4620	Minimum	S.F.	0.18	0.13	0.31
4640	Average	"	0.23	0.13	0.36
4660	Maximum	"	0.32	0.13	0.45
4680	Second Coat				
4700	Minimum	S.F.	0.13	0.13	0.26
4760	Average	"	0.16	0.13	0.29
4780	Maximum	"	0.21	0.13	0.34

		UNIT	LABOR	MAT.	TOTAL

09910.15 EXT. PAINTING, BUILDINGS

		UNIT	LABOR	MAT.	TOTAL
1200	Decks, Wood, Stained				
1580	Spray				
1600	First Coat				
1620	Minimum	S.F.	0.16	0.11	0.27
1640	Average	"	0.17	0.11	0.28
1660	Maximum	"	0.20	0.11	0.31
1680	Second Coat				
1700	Minimum	S.F.	0.14	0.11	0.25
1720	Average	"	0.15	0.11	0.26
1740	Maximum	"	0.17	0.11	0.28
2520	Doors, Wood				
2540	Brush				
2560	First Coat				
2580	Minimum	S.F.	0.80	0.13	0.93
2600	Average	"	1.04	0.13	1.17
2620	Maximum	"	1.30	0.13	1.43
2640	Second Coat				
2660	Minimum	S.F.	0.65	0.13	0.78
2680	Average	"	0.74	0.13	0.87
2700	Maximum	"	0.87	0.13	1.00
3680	Siding, Wood				
3880	Spray				
3900	First Coat				
3920	Minimum	S.F.	0.17	0.11	0.28
3940	Average	"	0.18	0.11	0.29
3960	Maximum	"	0.20	0.11	0.31
3980	Second Coat				
4000	Minimum	S.F.	0.13	0.11	0.24
4020	Average	"	0.17	0.11	0.28
4040	Maximum	"	0.26	0.11	0.37
4440	Trim				
4460	Brush				
4480	First Coat				
4500	Minimum	L.F.	0.21	0.16	0.37
4520	Average	"	0.26	0.16	0.42
4540	Maximum	"	0.32	0.16	0.48
4560	Second Coat				
4580	Minimum	L.F.	0.16	0.16	0.32
4600	Average	"	0.21	0.16	0.37
4620	Maximum	"	0.32	0.16	0.48
4640	Walls				
4840	Spray				
4860	First Coat				
4880	Minimum	S.F.	0.08	0.09	0.17
4900	Average	"	0.10	0.09	0.19
4920	Maximum	"	0.13	0.09	0.22
4940	Second Coat				
4960	Minimum	S.F.	0.06	0.09	0.15
4980	Average	"	0.08	0.09	0.17
5000	Maximum	"	0.11	0.09	0.20
5020	Windows				
5040	Brush				

		UNIT	LABOR	MAT.	TOTAL
09910.15	**EXT. PAINTING, BUILDINGS, Cont'd...**				
5060	First Coat				
5080	Minimum	S.F.	0.87	0.11	0.98
5100	Average	"	1.04	0.11	1.15
5120	Maximum	"	1.30	0.11	1.41
5140	Second Coat				
5160	Minimum	S.F.	0.74	0.11	0.85
5180	Average	"	0.87	0.11	0.98
5200	Maximum	"	1.04	0.11	1.15
09910.35	**INT. PAINTING, BUILDINGS**				
1380	Cabinets and Casework				
1400	Brush				
1420	First Coat				
1440	Minimum	S.F.	0.52	0.16	0.68
1460	Average	"	0.58	0.16	0.74
1480	Maximum	"	0.65	0.16	0.81
1500	Second Coat				
1520	Minimum	S.F.	0.43	0.16	0.59
1540	Average	"	0.47	0.16	0.63
1560	Maximum	"	0.52	0.16	0.68
1580	Spray				
1600	First Coat				
1620	Minimum	S.F.	0.26	0.13	0.39
1640	Average	"	0.30	0.13	0.43
1660	Maximum	"	0.37	0.13	0.50
1680	Second Coat				
1700	Minimum	S.F.	0.20	0.13	0.33
1720	Average	"	0.22	0.13	0.35
1740	Maximum	"	0.29	0.13	0.42
2520	Doors, Wood				
2540	Brush				
2560	First Coat				
2580	Minimum	S.F.	0.74	0.16	0.90
2600	Average	"	0.94	0.16	1.10
2620	Maximum	"	1.16	0.16	1.32
2640	Second Coat				
2660	Minimum	S.F.	0.58	0.12	0.70
2680	Average	"	0.65	0.12	0.77
2700	Maximum	"	0.74	0.12	0.86
2720	Spray				
2740	First Coat				
2760	Minimum	S.F.	0.15	0.12	0.27
2780	Average	"	0.18	0.12	0.30
2800	Maximum	"	0.23	0.12	0.35
2820	Second Coat				
2840	Minimum	S.F.	0.12	0.12	0.24
2860	Average	"	0.14	0.12	0.26
2880	Maximum	"	0.16	0.12	0.28
3900	Trim				
3920	Brush				
3940	First Coat				
3960	Minimum	L.F.	0.20	0.16	0.36

		UNIT	LABOR	MAT.	TOTAL
09910.35	**INT. PAINTING, BUILDINGS, Cont'd...**				
3980	Average	L.F.	0.23	0.16	0.39
4000	Maximum	"	0.29	0.16	0.45
4020	Second Coat				
4040	Minimum	L.F.	0.15	0.16	0.31
4060	Average	"	0.20	0.16	0.36
4080	Maximum	"	0.29	0.16	0.45
4100	Walls				
4120	Roller				
4140	First Coat				
4160	Minimum	S.F.	0.18	0.13	0.31
4180	Average	"	0.19	0.13	0.32
4200	Maximum	"	0.21	0.13	0.34
4220	Second Coat				
4240	Minimum	S.F.	0.16	0.13	0.29
4260	Average	"	0.17	0.13	0.30
4280	Maximum	"	0.20	0.13	0.33
4300	Spray				
4320	First Coat				
4340	Minimum	S.F.	0.08	0.11	0.19
4360	Average	"	0.10	0.11	0.21
4380	Maximum	"	0.13	0.11	0.24
4400	Second Coat				
4420	Minimum	S.F.	0.07	0.11	0.18
4440	Average	"	0.09	0.11	0.20
4460	Maximum	"	0.11	0.11	0.22
09955.10	**WALL COVERING**				
0900	Vinyl wall covering				
1000	Medium duty	S.F.	0.74	0.82	1.56
1010	Heavy duty	"	0.87	1.70	2.57

Design & Construction Resources

TABLE OF CONTENTS   PAGE

		UNIT	LABOR	MAT.	TOTAL
10110.10	**CHALKBOARDS**				
1020	Chalkboard, metal frame, 1/4" thick				
1040	48"x60"	EA.	52.00	370	422
1060	48"x96"	"	58.00	480	538
1080	48"x144"	"	65.00	830	895
1100	48"x192"	"	75.00	1,030	1,105
1110	Liquid chalkboard				
1120	48"x60"	EA.	52.00	400	452
1140	48"x96"	"	58.00	480	538
1160	48"x144"	"	65.00	830	895
1180	48"x192"	"	75.00	930	1,005
1200	Map rail, deluxe	L.F.	2.61	5.78	8.39
10165.10	**TOILET PARTITIONS**				
0100	Toilet partition, plastic laminate				
0120	Ceiling mounted	EA.	170	970	1,140
0140	Floor mounted	"	130	650	780
0150	Metal				
0165	Ceiling mounted	EA.	170	700	870
0180	Floor mounted	"	130	660	790
0190	Wheel chair partition, plastic laminate				
0200	Ceiling mounted	EA.	170	1,520	1,690
0210	Floor mounted	"	130	1,330	1,460
0215	Painted metal				
0220	Ceiling mounted	EA.	170	1,090	1,260
0230	Floor mounted	"	130	990	1,120
1980	Urinal screen, plastic laminate				
2000	Wall hung	EA.	65.00	470	535
2100	Floor mounted	"	65.00	420	485
2120	Porcelain enameled steel, floor mounted	"	65.00	540	605
2140	Painted metal, floor mounted	"	65.00	360	425
2160	Stainless steel, floor mounted	"	65.00	680	745
5000	Metal toilet partitions				
5020	Front door and side divider, floor mounted				
5040	Porcelain enameled steel	EA.	130	1,120	1,250
5060	Painted steel	"	130	650	780
5080	Stainless steel	"	130	1,930	2,060
10185.10	**SHOWER STALLS**				
1000	Shower receptors				
1010	Precast, terrazzo				
1020	32" x 32"	EA.	48.00	500	548
1040	32" x 48"	"	58.00	660	718
1050	Concrete				
1060	32" x 32"	EA.	48.00	250	298
1080	48" x 48"	"	64.00	290	354
1100	Shower door, trim and hardware				
1130	Porcelain enameled steel, flush	EA.	58.00	500	558
1140	Baked enameled steel, flush	"	58.00	290	348
1150	Aluminum frame, tempered glass, 48" wide, sliding	"	72.00	620	692
1161	Folding	"	72.00	590	662
5400	Shower compartment, precast concrete receptor				
5420	Single entry type				
5440	Porcelain enameled steel	EA.	580	2,360	2,940

		UNIT	LABOR	MAT.	TOTAL
10185.10	**SHOWER STALLS, Cont'd...**				
5460	Baked enameled steel	EA.	580	1,980	2,560
5480	Stainless steel	"	580	1,930	2,510
5500	Double entry type				
5520	Porcelain enameled steel	EA.	720	3,960	4,680
5540	Baked enameled steel	"	720	3,010	3,730
5560	Stainless steel	"	720	4,400	5,120
10290.10	**PEST CONTROL**				
1000	Termite control				
1010	Under slab spraying				
1020	Minimum	S.F.	0.10	1.07	1.17
1040	Average	"	0.20	1.10	1.30
1120	Maximum	"	0.40	1.48	1.88
10350.10	**FLAGPOLES**				
2020	Installed in concrete base				
2030	Fiberglass				
2040	25' high	EA.	350	1,230	1,580
2080	50' high	"	870	4,480	5,350
2100	Aluminum				
2120	25' high	EA.	350	1,270	1,620
2140	50' high	"	870	3,840	4,710
2160	Bonderized steel				
2180	25' high	EA.	400	1,420	1,820
2200	50' high	"	1,040	2,840	3,880
2220	Freestanding tapered, fiberglass				
2240	30' high	EA.	370	1,530	1,900
2260	40' high	"	470	2,140	2,610
2280	50' high	"	520	5,050	5,570
2300	60' high	"	610	5,860	6,470
2400	Wall mounted, with collar, brushed aluminum finish				
2420	15' long	EA.	260	990	1,250
2440	18' long	"	260	1,190	1,450
2460	20' long	"	270	1,380	1,650
2480	24' long	"	310	1,400	1,710
2500	Outrigger, wall, including base				
2520	10' long	EA.	350	1,250	1,600
2540	20' long	"	430	1,560	1,990
10400.10	**IDENTIFYING DEVICES**				
1000	Directory and bulletin boards				
1020	Open face boards				
1040	Chrome plated steel frame	S.F.	26.00	28.75	54.75
1060	Aluminum framed	"	26.00	49.25	75.25
1080	Bronze framed	"	26.00	64.00	90.00
1100	Stainless steel framed	"	26.00	89.00	115
1140	Tack board, aluminum framed	"	26.00	20.25	46.25
1160	Visual aid board, aluminum framed	"	26.00	20.25	46.25
1200	Glass encased boards, hinged and keyed				
1210	Aluminum framed	S.F.	65.00	110	175
1220	Bronze framed	"	65.00	120	185
1230	Stainless steel framed	"	65.00	160	225
1240	Chrome plated steel framed	"	65.00	170	235
2020	Metal plaque				

10400.10	IDENTIFYING DEVICES, Cont'd...	UNIT	LABOR	MAT.	TOTAL
2040	Cast bronze	S.F.	43.50	480	524
2060	Aluminum	"	43.50	280	324
2080	Metal engraved plaque				
2100	Porcelain steel	S.F.	43.50	600	644
2120	Stainless steel	"	43.50	480	524
2140	Brass	"	43.50	720	764
2160	Aluminum	"	43.50	430	474
2200	Metal built-up plaque				
2220	Bronze	S.F.	52.00	540	592
2240	Copper and bronze	"	52.00	470	522
2260	Copper and aluminum	"	52.00	520	572
2280	Metal nameplate plaques				
2300	Cast bronze	S.F.	32.50	530	563
2320	Cast aluminum	"	32.50	390	423
2330	Engraved, 1-1/2" x 6"				
2340	Bronze	EA.	32.50	220	253
2360	Aluminum	"	32.50	170	203
2440	Letters, on masonry or concrete, aluminum, satin finish				
2450	1/2" thick				
2460	2" high	EA.	20.75	21.25	42.00
2480	4" high	"	26.00	31.75	57.75
2500	6" high	"	29.00	42.50	71.50
2510	3/4" thick				
2520	8" high	EA.	32.50	64.00	96.50
2540	10" high	"	37.25	73.00	110
2550	1" thick				
2560	12" high	EA.	43.50	86.00	130
2580	14" high	"	52.00	99.00	151
2600	16" high	"	65.00	120	185
2620	For polished aluminum add, 15%				
2640	For clear anodized aluminum add, 15%				
2660	For colored anodic aluminum add, 30%				
2680	For profiled and color enameled letters add, 50%				
2700	Cast bronze, satin finish letters				
2710	3/8" thick				
2720	2" high	EA.	20.75	24.25	45.00
2740	4" high	"	26.00	36.25	62.25
2760	1/2" thick, 6" high	"	29.00	51.00	80.00
2780	5/8" thick, 8" high	"	32.50	74.00	107
2785	1" thick				
2790	10" high	EA.	37.25	90.00	127
2800	12" high	"	43.50	110	154
2820	14" high	"	52.00	140	192
2840	16" high	"	65.00	200	265
3000	Interior door signs, adhesive, flexible				
3060	2" x 8"	EA.	10.25	19.75	30.00
3080	4" x 4"	"	10.25	21.00	31.25
3100	6" x 7"	"	10.25	26.50	36.75
3120	6" x 9"	"	10.25	34.00	44.25
3140	10" x 9"	"	10.25	45.00	55.25
3160	10" x 12"	"	10.25	58.00	68.25
3200	Hard plastic type, no frame				

		UNIT	LABOR	MAT.	TOTAL
10400.10	**IDENTIFYING DEVICES, Cont'd...**				
3220	3" x 8"	EA.	10.25	46.25	56.50
3240	4" x 4"	"	10.25	44.00	54.25
3260	4" x 12"	"	10.25	48.50	58.75
3280	Hard plastic type, with frame				
3300	3" x 8"	EA.	10.25	140	150
3320	4" x 4"	"	10.25	110	120
3340	4" x 12"	"	10.25	170	180
10450.10	**CONTROL**				
1020	Access control, 7' high, indoor or outdoor impenetrability				
1040	Remote or card control, type B	EA.	710	1,520	2,230
1060	Free passage, type B	"	710	1,210	1,920
1080	Remote or card control, type AA	"	710	2,420	3,130
1100	Free passage, type AA	"	710	2,260	2,970
10500.10	**LOCKERS**				
0080	Locker bench, floor mounted, laminated maple				
0100	4'	EA.	43.50	250	294
0120	6'	"	43.50	350	394
0130	Wardrobe locker, 12" x 60" x 15", baked on enamel				
0140	1-tier	EA.	26.00	270	296
0160	2-tier	"	26.00	280	306
0180	3-tier	"	27.50	320	348
0200	4-tier	"	27.50	340	368
0240	12" x 72" x 15", baked on enamel				
0260	1-tier	EA.	26.00	230	256
0280	2-tier	"	26.00	280	306
0300	4-tier	"	27.50	340	368
0320	5-tier	"	27.50	340	368
1200	15" x 60" x 15", baked on enamel				
1220	1-tier	EA.	26.00	300	326
1240	4-tier	"	27.50	330	358
10520.10	**FIRE PROTECTION**				
1000	Portable fire extinguishers				
1020	Water pump tank type				
1030	2.5 gal.				
1040	Red enameled galvanized	EA.	27.00	100	127
1060	Red enameled copper	"	27.00	160	187
1080	Polished copper	"	27.00	220	247
1200	Carbon dioxide type, red enamel steel				
1210	Squeeze grip with hose and horn				
1220	2.5 lb	EA.	27.00	120	147
1240	5 lb	"	31.25	180	211
1260	10 lb	"	40.50	360	401
1280	15 lb	"	51.00	290	341
1300	20 lb	"	51.00	360	411
1310	Wheeled type				
1320	125 lb	EA.	81.00	2,360	2,441
1340	250 lb	"	81.00	3,410	3,491
1360	500 lb	"	81.00	4,610	4,691
1400	Dry chemical, pressurized type				
1405	Red enameled steel				
1410	2.5 lb	EA.	27.00	47.00	74.00

DIVISION # 10 SPECIALTIES

		UNIT	LABOR	MAT.	TOTAL
10520.10	**FIRE PROTECTION, Cont'd...**				
1430	5 lb	EA.	31.25	74.00	105
1440	10 lb	"	40.50	84.00	125
1450	20 lb	"	51.00	120	171
1460	30 lb	"	51.00	160	211
1480	Chrome plated steel, 2.5 lb	"	27.00	150	177
10550.10	**POSTAL SPECIALTIES**				
1500	Mail chutes				
1520	Single mail chute				
1530	Finished aluminum	L.F.	130	690	820
1540	Bronze	"	130	1,000	1,130
1560	Single mail chute receiving box				
1580	Finished aluminum	EA.	260	1,070	1,330
1600	Bronze	"	260	1,260	1,520
1620	Twin mail chute, double parallel				
1630	Finished aluminum	FLR	260	1,310	1,570
1640	Bronze	"	260	2,060	2,320
10670.10	**SHELVING**				
0980	Shelving, enamel, closed side and back, 12" x 36"				
1000	5 shelves	EA.	87.00	250	337
1020	8 shelves	"	120	270	390
1030	Open				
1040	5 shelves	EA.	87.00	130	217
1060	8 shelves	"	120	140	260
2000	Metal storage shelving, baked enamel				
2030	7 shelf unit, 72" or 84" high				
2050	12" shelf	L.F.	55.00	40.50	95.50
2080	24" shelf	"	65.00	82.00	147
2100	36" shelf	"	75.00	95.00	170
2200	4 shelf unit, 40" high				
2240	12" shelf	L.F.	47.50	55.00	103
2270	24" shelf	"	58.00	110	168
2300	3 shelf unit, 32" high				
2340	12" shelf	L.F.	27.50	41.75	69.25
2370	24" shelf	"	32.50	54.00	86.50
2400	Single shelf unit, attached to masonry				
2420	12" shelf	L.F.	9.49	16.25	25.74
2450	24" shelf	"	11.25	23.25	34.50
2460	For stainless steel, add to material, 120%				
2470	For attachment to gypsum board, add to labor, 50%				
10800.10	**BATH ACCESSORIES**				
1040	Ash receiver, wall mounted, aluminum	EA.	26.00	130	156
1050	Grab bar, 1-1/2" dia., stainless steel, wall mounted				
1060	24" long	EA.	26.00	69.00	95.00
1080	36" long	"	27.50	80.00	108
1100	48" long	"	30.75	98.00	129
1130	1" dia., stainless steel				
1140	12" long	EA.	22.75	43.00	65.75
1180	24" long	"	26.00	51.00	77.00
1220	36" long	"	29.00	59.00	88.00
1240	48" long	"	30.75	67.00	97.75
1300	Hand dryer, surface mounted, 110 volt	"	65.00	600	665

		UNIT	LABOR	MAT.	TOTAL
10800.10	**BATH ACCESSORIES, Cont'd...**				
1320	Medicine cabinet, 16 x 22, baked enamel, steel, lighted	EA.	20.75	110	131
1340	With mirror, lighted	"	34.75	190	225
1420	Mirror, 1/4" plate glass, up to 10 sf	S.F.	5.22	9.90	15.12
1430	Mirror, stainless steel frame				
1440	18"x24"	EA.	17.50	76.00	93.50
1500	24"x30"	"	26.00	92.00	118
1520	24"x48"	"	43.50	140	184
1560	30"x30"	"	52.00	300	352
1600	48"x72"	"	87.00	590	677
1640	With shelf, 18"x24"	"	20.75	230	251
1820	Sanitary napkin dispenser, stainless steel, wall mounted	"	34.75	560	595
1830	Shower rod, 1" diameter				
1840	Chrome finish over brass	EA.	26.00	200	226
1860	Stainless steel	"	26.00	140	166
1900	Soap dish, stainless steel, wall mounted	"	34.75	130	165
1910	Toilet tissue dispenser, stainless, wall mounted				
1920	Single roll	EA.	13.00	60.00	73.00
1940	Double roll	"	15.00	120	135
1945	Towel dispenser, stainless steel				
1950	Flush mounted	EA.	29.00	230	259
1960	Surface mounted	"	26.00	330	356
1970	Combination towel dispenser and waste receptacle	"	34.75	500	535
2000	Towel bar, stainless steel				
2020	18" long	EA.	20.75	75.00	95.75
2040	24" long	"	23.75	100	124
2060	30" long	"	26.00	110	136
2070	36" long	"	29.00	120	149
2100	Waste receptacle, stainless steel, wall mounted	"	43.50	380	424

Design & Construction Resources

TABLE OF CONTENTS   **PAGE**

		UNIT	LABOR	MAT.	TOTAL
11010.10	**MAINTENANCE EQUIPMENT**				
1000	Vacuum cleaning system				
1010	3 valves				
1020	1.5 hp	EA.	580	2,010	2,590
1030	2.5 hp	"	750	2,110	2,860
1040	5 valves	"	1,040	2,530	3,570
1060	7 valves	"	1,310	3,250	4,560
11060.10	**THEATER EQUIPMENT**				
1000	Roll out stage, steel frame, wood floor				
1020	Manual	S.F.	3.26	42.25	45.51
1040	Electric	"	5.22	46.00	51.22
1100	Portable stages				
1120	8" high	S.F.	2.61	23.00	25.61
1140	18" high	"	2.90	25.00	27.90
1160	36" high	"	3.07	26.00	29.07
1180	48" high	"	3.26	37.50	40.76
1300	Band risers				
1320	Minimum	S.F.	2.61	27.50	30.11
1340	Maximum	"	2.61	59.00	61.61
1400	Chairs for risers				
1420	Minimum	EA.	1.84	800	802
1440	Maximum	"	1.84	1,130	1,132
2000	Theatre controlls				
2010	Fade console, 48 channel	EA.	270	2,200	2,470
2020	Light control modules, 125 channels	"	530	7,370	7,900
2030	Dimmer module, stage-pin output	"	270	2,150	2,420
2040	6-Module pack, w/24 U-ground connectors	"	270	6,060	6,330
11090.10	**CHECKROOM EQUIPMENT**				
1000	Motorized checkroom equipment				
1020	No shelf system, 6'4" height				
1040	7'6" length	EA.	520	3,010	3,530
1060	14'6" length	"	520	3,470	3,990
1080	28' length	"	520	4,400	4,920
1100	One shelf, 6'8" height				
1120	7'6" length	EA.	520	3,650	4,170
1140	14'6" length	"	520	4,420	4,940
1160	28' length	"	520	6,540	7,060
11110.10	**LAUNDRY EQUIPMENT**				
1000	High capacity, heavy duty				
1020	Washer extractors				
1030	135 lb				
1040	Standard	EA.	430	60,500	60,930
1060	Pass through	"	430	70,950	71,380
1070	200 lb				
1080	Standard	EA.	430	68,200	68,630
1100	Pass through	"	430	76,450	76,880
1120	110 lb dryer	"	430	12,020	12,450
11161.10	**LOADING DOCK EQUIPMENT**				
0080	Dock leveler, 10 ton capacity				
0100	6' x 8'	EA.	520	5,240	5,760
0120	7' x 8'	"	520	5,530	6,050

		UNIT	LABOR	MAT.	TOTAL
11170.10	**WASTE HANDLING**				
1500	Industrial compactor				
1520	1 cy	EA.	590	12,100	12,690
1540	3 cy	"	760	23,720	24,480
1560	5 cy	"	1,070	30,800	31,870
11170.20	**AERATION EQUIPMENT**				
0010	Surface spray/Vertical pump				
0020	1 hp. Pump, 500 gpm.	EA.	290	5,660	5,950
0030	5 hp. Pump, 2000 gpm.	"	1,150	7,960	9,110
0040	Polycarbon, treatment container, 1,000 gallon capacity	"	2,230	4,520	6,750
0050	1,500 gallon	"	2,230	9,430	11,660
0060	2,000 gallon	"	2,790	12,730	15,520
0070	3,000 gallon	"	2,970	17,130	20,100
11400.10	**FOOD SERVICE EQUIPMENT**				
1000	Unit kitchens				
1020	30" compact kitchen				
1040	Refrigerator, with range, sink	EA.	270	1,750	2,020
1060	Sink only	"	180	1,630	1,810
1080	Range only	"	130	1,330	1,460
1100	Cabinet for upper wall section	"	76.00	330	406
1120	Stainless shield, for rear wall	"	21.25	130	151
1140	Side wall	"	21.25	98.00	119
1200	42" compact kitchen				
1220	Refrigerator with range, sink	EA.	300	1,970	2,270
1240	Sink only	"	270	1,960	2,230
1260	Cabinet for upper wall section	"	89.00	660	749
1280	Stainless shield, for rear wall	"	22.25	410	432
1290	Side wall	"	22.25	150	172
1600	Bake oven				
1610	Single deck				
1620	Minimum	EA.	67.00	3,030	3,097
1640	Maximum	"	130	5,780	5,910
1650	Double deck				
1660	Minimum	EA.	89.00	5,390	5,479
1680	Maximum	"	130	16,840	16,970
1685	Triple deck				
1690	Minimum	EA.	89.00	19,110	19,199
1695	Maximum	"	180	34,100	34,280
1700	Convection type oven, electric, 40" x 45" x 57"				
1720	Minimum	EA.	67.00	4,550	4,617
1740	Maximum	"	130	8,000	8,130
1800	Broiler, without oven, 69" x 26" x 39"				
1820	Minimum	EA.	67.00	4,020	4,087
1840	Maximum	"	89.00	4,950	5,039
1900	Coffee urns, 10 gallons				
1920	Minimum	EA.	180	2,590	2,770
1940	Maximum	"	270	4,450	4,720
2000	Fryer, with submerger				
2010	Single				
2020	Minimum	EA.	110	1,500	1,610
2040	Maximum	"	180	14,530	14,710
2050	Double				

11400.10 FOOD SERVICE EQUIPMENT, Cont'd...	UNIT	LABOR	MAT.	TOTAL
2060 Minimum	EA.	130	2,090	2,220
2080 Maximum	"	180	15,730	15,910
2100 Griddle, counter				
2110 3' long				
2120 Minimum	EA.	89.00	1,900	1,989
2140 Maximum	"	110	5,170	5,280
2150 5' long				
2160 Minimum	EA.	130	4,700	4,830
2180 Maximum	"	180	7,900	8,080
2200 Kettles, steam, jacketed				
2210 20 gallons				
2220 Minimum	EA.	130	10,070	10,200
2240 Maximum	"	270	14,480	14,750
2300 Range				
2310 Heavy duty, single oven, open top				
2320 Minimum	EA.	67.00	5,890	5,957
2340 Maximum	"	180	11,000	11,180
2350 Fry top				
2360 Minimum	EA.	67.00	6,120	6,187
2380 Maximum	"	180	8,750	8,930
2400 Steamers, electric				
2410 27 kw				
2420 Minimum	EA.	130	10,910	11,040
2440 Maximum	"	180	20,780	20,960
2450 18 kw				
2460 Minimum	EA.	130	5,920	6,050
2480 Maximum	"	180	14,300	14,480
2500 Dishwasher, rack type				
2520 Single tank, 190 racks/hr	EA.	270	16,040	16,310
2530 Double tank				
2540 234 racks/hr	EA.	300	31,820	32,120
2560 265 racks/hr	"	360	45,700	46,060
2580 Dishwasher, automatic 100 meals/hr	"	180	12,700	12,880
2600 Disposals				
2620 100 gal/hr	EA.	180	1,060	1,240
2640 120 gal/hr	"	180	1,270	1,450
2660 250 gal/hr	"	190	1,500	1,690
2680 Exhaust hood for dishwasher, gutter 4 sides, s-steel				
2681 4'x4'x2'	EA.	200	2,450	2,650
2690 4'x7'x2'	"	210	3,560	3,770
2900 Ice cube maker				
2910 50 lb per day				
2920 Minimum	EA.	530	1,850	2,380
2940 Maximum	"	530	2,910	3,440
2950 500 lb per day				
2960 Minimum	EA.	890	4,550	5,440
2970 Maximum	"	890	5,060	5,950
3100 Refrigerated cases				
3120 Dairy products				
3140 Multi deck type	L.F.	35.50	1,080	1,116
3160 For rear sliding doors, add	"			120
3180 Delicatessen case, service deli				

		UNIT	LABOR	MAT.	TOTAL
11400.10	**FOOD SERVICE EQUIPMENT, Cont'd...**				
3190	Single deck	L.F.	270	690	960
3200	Multi deck	"	330	810	1,140
3220	Meat case				
3230	Single deck	L.F.	310	620	930
3240	Multi deck	"	330	720	1,050
3260	Produce case				
3270	Single deck	L.F.	310	680	990
3280	Multi deck	"	330	770	1,100
3300	Bottle coolers				
3310	6' long				
3320	Minimum	EA.	1,070	2,510	3,580
3340	Maximum	"	1,070	3,850	4,920
3345	10' long				
3350	Minimum	EA.	1,780	3,060	4,840
3360	Maximum	"	1,780	5,850	7,630
3420	Frozen food cases				
3440	Chest type	L.F.	310	520	830
3460	Reach-in, glass door	"	330	750	1,080
3470	Island case, single	"	310	660	970
3480	Multi deck	"	330	1,050	1,380
3500	Ice storage bins				
3520	500 lb capacity	EA.	760	1,830	2,590
3530	1000 lb capacity	"	1,530	3,600	5,130
11450.10	**RESIDENTIAL EQUIPMENT**				
0300	Compactor, 4 to 1 compaction	EA.	130	790	920
1310	Dishwasher, built-in				
1320	2 cycles	EA.	270	650	920
1330	4 or more cycles	"	270	2,170	2,440
1340	Disposal				
1350	Garbage disposer	EA.	180	150	330
1360	Heaters, electric, built-in				
1362	Ceiling type	EA.	180	400	580
1364	Wall type				
1370	Minimum	EA.	130	210	340
1374	Maximum	"	180	470	650
1400	Hood for range, 2-speed, vented				
1420	30" wide	EA.	180	420	600
1440	42" wide	"	180	910	1,090
1460	Ice maker, automatic				
1480	30 lb per day	EA.	76.00	1,540	1,616
1500	50 lb per day	"	270	2,310	2,580
2000	Ranges electric				
2040	Built-in, 30", 1 oven	EA.	180	1,890	2,070
2050	2 oven	"	180	2,120	2,300
2060	Counter top, 4 burner, standard	"	130	1,040	1,170
2070	With grill	"	130	2,760	2,890
2198	Free standing, 21", 1 oven	"	180	1,050	1,230
2200	30", 1 oven	"	110	1,760	1,870
2220	2 oven	"	110	3,490	3,600
3600	Water softener				
3620	30 grains per gallon	EA.	180	1,010	1,190

		UNIT	LABOR	MAT.	TOTAL
11450.10	**RESIDENTIAL EQUIPMENT, Cont'd...**				
3640	70 grains per gallon	EA.	270	1,270	1,540
11600.10	**LABORATORY EQUIPMENT**				
1000	Cabinets, base				
1020	Minimum	L.F.	43.50	380	424
1040	Maximum	"	43.50	640	684
1080	Full storage, 7' high				
1100	Minimum	L.F.	43.50	480	524
1140	Maximum	"	43.50	660	704
1150	Wall				
1160	Minimum	L.F.	52.00	140	192
1200	Maximum	"	52.00	240	292
1220	Counter tops				
1240	Minimum	S.F.	6.52	30.75	37.27
1260	Average	"	7.45	77.00	84.45
1280	Maximum	"	8.70	110	119
1300	Tables				
1320	Open underneath	S.F.	26.00	94.00	120
1330	Doors underneath	"	32.50	370	403
2000	Medical laboratory equipment				
2010	Analyzer				
2020	Chloride	EA.	26.75	870	897
2060	Blood	"	44.50	32,080	32,125
2070	Bath, water, utility, countertop unit	"	53.00	810	863
2080	Hot plate, lab, countertop	"	48.50	350	399
2100	Stirrer	"	48.50	390	439
2120	Incubator, anaerobic, 23x23x36"	"	270	8,340	8,610
2140	Dry heat bath	"	89.00	680	769
2160	Incinerator, for sterilizing	"	5.33	320	325
2170	Meter, serum protein	"	6.67	1,360	1,367
2180	Ph analog, general purpose	"	7.62	1,620	1,628
2190	Refrigerator, blood bank, undercounter type 153 litres	"	89.00	6,020	6,109
2200	5.4 cf, undercounter type	"	89.00	4,960	5,049
2210	Refrigerator/freezer, 4.4 cf, undercounter type	"	89.00	770	859
2220	Sealer, impulse, free standing, 20x12x4"	"	17.75	370	388
2240	Timer, electric, 1-60 minutes, bench or wall mounted	"	29.75	180	210
2260	Glassware washer - dryer, undercounter	"	670	7,820	8,490
2300	Balance, torsion suspension, tabletop, 4.5 lb capacity	"	29.75	710	740
2340	Binocular microscope, with in base illuminator	"	20.50	3,420	3,441
2400	Centrifuge, table model, 19x16x13"	"	21.25	2,270	2,291
2420	Clinical model, with four place head	"	11.75	1,030	1,042
11700.10	**MEDICAL EQUIPMENT**				
1000	Hospital equipment, lights				
1020	Examination, portable	EA.	44.50	1,860	1,905
1200	Meters				
1220	Air flow meter	EA.	29.75	82.00	112
1240	Oxygen flow meters	"	22.25	64.00	86.25
1900	Physical therapy				
1930	Chair, hydrotherapy	EA.	8.70	540	549
1940	Diathermy, shortwave, portable, on casters	"	20.75	9,210	9,231
1950	Exercise bicycle, floor standing, 35" x 15"	"	17.50	1,080	1,098
1960	Hydrocollator, 4 pack, portable, 129 x 90 x 160"	"	7.45	500	507

		UNIT	LABOR	MAT.	TOTAL
11700.10	**MEDICAL EQUIPMENT, Cont'd...**				
1970	Lamp, infrared, mobile with variable heat control	EA.	40.25	510	550
1980	Ultra violet, base mounted	"	40.25	590	630
2070	Stimulator, galvanic-faradic, hand held	"	3.48	310	313
2080	Ultrasound, muscle stimulator, portable, 13x13x8"	"	4.35	2,200	2,204
2120	Whirlpool, 85 gallon	"	260	4,400	4,660
2141	65 gallon capacity	"	260	4,070	4,330
2200	Radiology				
2280	Radiographic table, motor driven tilting table	EA.	5,220	38,060	43,280
2290	Fluoroscope image/tv system	"	10,440	100,980	111,420
2300	Processor for washing and drying radiographs				
2303	Water filter unit, 30" x 48-1/2" x 37-1/2"	EA.	890	94.00	984
2400	Steam sterilizers				
2410	For heat and moisture stable materials	EA.	53.00	4,750	4,803
2440	For fast drying after sterilization	"	67.00	6,090	6,157
2450	Compact unit	"	67.00	2,010	2,077
2460	Semi-automatic	"	270	7,700	7,970
2465	Floor loading				
2470	Single door	EA.	440	65,670	66,110
2480	Double door	"	530	71,500	72,030
2490	Utensil washer, sanitizer	"	410	13,200	13,610
2500	Automatic washer/sterilizer	"	1,070	8,360	9,430
2510	16 x 16 x 26", including generator & accessories	"	1,780	9,130	10,910
2520	Steam generator, elec., 10 kw to 180 kw	"	1,070	21,230	22,300
2540	Surgical scrub				
2560	Minimum	EA.	180	1,470	1,650
2580	Maximum	"	180	8,340	8,520
2600	Gas sterilizers				
2620	Automatic, free standing, 21x19x29"	EA.	530	5,280	5,810
2720	Surgical lights, ceiling mounted				
2740	Minimum	EA.	890	7,550	8,440
2760	Maximum	"	1,070	14,880	15,950
2800	Water stills				
2900	4 liters/hr	EA.	180	3,530	3,710
2920	8 liters/hr	"	180	6,320	6,500
2940	19 liters/hr	"	440	11,580	12,020
3060	X-ray equipment				
3070	Mobile unit				
3080	Minimum	EA.	270	6,490	6,760
3100	Maximum	"	530	13,090	13,620
3220	Autopsy Table, Minimum	"			11,420
3240	Maximum	"			26,520
3300	Incubators				
3320	15 cf	EA.	270	6,310	6,580
3330	29 cf	"	440	6,900	7,340
3340	Infant transport, portable	"	280	5,500	5,780
3460	Headwall				
3465	Aluminum, with back frame and console	EA.	270	4,290	4,560
6000	Hospital ground detection system				
6010	Power ground module	EA.	150	1,070	1,220
6020	Ground slave module	"	120	700	820
6030	Master ground module	"	100	440	540
6040	Remote indicator	"	110	480	590

11700.10	MEDICAL EQUIPMENT, Cont'd...	UNIT	LABOR	MAT.	TOTAL
6050	X-ray indicator	EA.	120	1,380	1,500
6060	Micro ammeter	"	130	1,720	1,850
6070	Supervisory module	"	120	1,390	1,510
6080	Ground cords	"	19.75	120	140
6100	Hospital isolation monitors, 5 ma				
6110	120v	EA.	230	2,950	3,180
6120	208v	"	230	2,640	2,870
6130	240v	"	230	2,640	2,870
6210	Digital clock-timers separate display	"	110	1,070	1,180
6220	One display	"	110	1,030	1,140
6230	Remote control	"	83.00	600	683
6240	Battery pack	"	83.00	140	223
6310	Surgical chronometer clock and 3 timers	"	170	2,270	2,440
6320	Auxilary control	"	77.00	820	897

TABLE OF CONTENTS   PAGE

		UNIT	LABOR	MAT.	TOTAL
12302.10	**CASEWORK**				
0080	Kitchen base cabinet, prefinished, 24" deep, 35" high				
0100	12"wide	EA.	52.00	210	262
0120	18" wide	"	52.00	270	322
0140	24" wide	"	58.00	280	338
0160	27" wide	"	58.00	330	388
0180	36" wide	"	65.00	380	445
0200	48" wide	"	65.00	430	495
0220	Corner cabinet, 36" wide	"	65.00	450	515
4000	Wall cabinet, 12" deep, 12" high				
4020	30" wide	EA.	52.00	180	232
4060	36" wide	"	52.00	220	272
4110	24" high				
4120	30" wide	EA.	58.00	240	298
4140	36" wide	"	58.00	270	328
4150	30" high				
4160	12" wide	EA.	65.00	160	225
4200	24" wide	"	65.00	190	255
4320	30" wide	"	75.00	260	335
4340	36" wide	"	75.00	260	335
4350	Corner cabinet, 30" high				
4360	24" wide	EA.	87.00	250	337
4390	36" wide	"	87.00	340	427
5020	Wardrobe	"	130	770	900
6980	Vanity with top, laminated plastic				
7000	24" wide	EA.	130	480	610
7040	36" wide	"	170	610	780
7060	48" wide	"	210	700	910
12390.10	**COUNTER TOPS**				
1020	Stainless steel, counter top, with backsplash	S.F.	13.00	120	133
2000	Acid-proof, kemrock surface	"	8.70	41.75	50.45
12500.10	**WINDOW TREATMENT**				
1000	Drapery tracks, wall or ceiling mounted				
1040	Basic traverse rod				
1080	50 to 90"	EA.	26.00	49.75	75.75
1100	84 to 156"	"	29.00	75.00	104
1120	136 to 250"	"	29.00	95.00	124
1140	165 to 312"	"	32.50	140	173
1160	Traverse rod with stationary curtain rod				
1180	30 to 50"	EA.	26.00	67.00	93.00
1200	50 to 90"	"	26.00	76.00	102
1220	84 to 156"	"	29.00	110	139
1240	136 to 250"	"	32.50	130	163
1260	Double traverse rod				
1280	30 to 50"	EA.	26.00	76.00	102
1300	50 to 84"	"	26.00	97.00	123
1320	84 to 156"	"	29.00	120	149
1340	136 to 250"	"	32.50	140	173
12510.10	**BLINDS**				
0990	Venetian blinds				
1000	2" slats	S.F.	1.30	29.75	31.05
1020	1" slats	"	1.30	31.50	32.80

TABLE OF CONTENTS   PAGE

		UNIT	LABOR	MAT.	TOTAL
13056.10	**VAULTS**				
1000	Floor safes				
1010	1.0 cf	EA.	43.50	780	824
1020	1.3 cf	"	65.00	860	925
13121.10	**PRE-ENGINEERED BUILDINGS**				
1080	Pre-engineered metal building, 40'x100'				
1100	14' eave height	S.F.	4.45	6.60	11.05
1120	16' eave height	"	5.14	7.46	12.60
1140	20' eave height	"	6.68	8.42	15.10
1150	60'x100'				
1160	14' eave height	S.F.	4.45	8.40	12.85
1180	16' eave height	"	5.14	9.17	14.31
1190	20' eave height	"	6.68	10.25	16.93
1195	80'x100'				
1200	14' eave height	S.F.	4.45	6.38	10.83
1210	16' eave height	"	5.14	6.62	11.76
1220	20' eave height	"	6.68	7.37	14.05
1280	100'x100'				
1300	14' eave height	S.F.	4.45	6.30	10.75
1320	16' eave height	"	5.14	6.50	11.64
1340	20' eave height	"	6.68	7.15	13.83
1350	100'x150'				
1360	14' eave height	S.F.	4.45	5.50	9.95
1380	16' eave height	"	5.14	5.76	10.90
1400	20' eave height	"	6.68	6.13	12.81
1410	120'x150'				
1420	14' eave height	S.F.	4.45	5.85	10.30
1440	16' eave height	"	5.14	5.99	11.13
1460	20' eave height	"	6.68	6.32	13.00
1480	140'x150'				
1500	14' eave height	S.F.	4.45	5.50	9.95
1520	16' eave height	"	5.14	5.68	10.82
1540	20' eave height	"	6.68	6.13	12.81
1600	160'x200'				
1620	14' eave height	S.F.	4.45	4.26	8.71
1640	16' eave height	"	5.14	4.40	9.54
1680	20' eave height	"	6.68	4.67	11.35
1690	200'x200'				
1700	14' eave height	S.F.	4.45	3.68	8.13
1720	16' eave height	"	5.14	4.04	9.18
1740	20' eave height	"	6.68	4.26	10.94
5020	Hollow metal door and frame, 6' x 7'	EA.			980
5030	Sectional steel overhead door, manually operated				
5040	8' x 8'	EA.			1,610
5080	12' x 12'	"			2,140
5100	Roll-up steel door, manually operated				
5120	10' x 10'	EA.			1,250
5140	12' x 12'	"			2,250
5160	For gravity ridge ventilator with birdscreen	"			540
5161	9" throat x 10'	"			590
5181	12" throat x 10'	"			720
5200	For 20" rotary vent with damper	"			270

		UNIT	LABOR	MAT.	TOTAL
13121.10	**PRE-ENGINEERED BUILDINGS, Cont'd...**				
5220	For 4' x 3' fixed louver	EA.			190
5240	For 4' x 3' aluminum sliding window	"			170
5260	For 3' x 9' fiberglass panels	"			130
8020	Liner panel, 26 ga, painted steel	S.F.	1.45	2.36	3.81
8040	Wall panel insulated, 26 ga. steel, foam core	"	1.45	7.42	8.87
8060	Roof panel, 26 ga. painted steel	"	0.83	2.20	3.03
8080	Plastic (sky light)	"	0.83	5.00	5.83
9000	Insulation, 3-1/2" thick blanket, R11	"	0.38	1.48	1.86
13152.10	**SWIMMING POOL EQUIPMENT**				
1100	Diving boards				
1110	14' long				
1120	Aluminum	EA.	230	3,830	4,060
1140	Fiberglass	"	230	2,740	2,970
1500	Ladders, heavy duty				
1510	2 steps				
1520	Minimum	EA.	81.00	1,090	1,171
1540	Maximum	"	81.00	1,320	1,401
1550	4 steps				
1560	Minimum	EA.	100	1,370	1,470
1580	Maximum	"	100	1,530	1,630
1600	Lifeguard chair				
1620	Minimum	EA.	410	3,410	3,820
1640	Maximum	"	410	4,040	4,450
1700	Lights, underwater				
1720	12 volt, with transformer	EA.	100	500	600
1730	110 volt				
1740	Minimum	EA.	100	630	730
1760	Maximum	"	100	1,420	1,520
1780	Ground fault interrupter for 110 volt, each light	"	33.75	190	224
2000	Pool covers				
2020	Reinforced polyethylene	S.F.	3.12	1.92	5.04
2030	Vinyl water tube				
2040	Minimum	S.F.	3.12	1.10	4.22
2060	Maximum	"	3.12	1.65	4.77
2100	Slides with water tube				
2120	Minimum	EA.	340	700	1,040
2140	Maximum	"	340	14,300	14,640
13152.20	**SAUNAS**				
0010	Prefabricated, cedar siding, insulated panels, prehung door,				
0020	4'x8"x4'-8"x6'-6"	EA.			4,650
0030	5'-8"x6'-8"x6'-6"	"			5,660
0040	6'-8"x6'-8"x6'-6"	"			6,540
0050	7'-8"x7'-8"x6'-6"	"			7,950
0060	7'-8"x9'-8"x6'-6"	"			10,600
13200.10	**STORAGE TANKS**				
0080	Oil storage tank, underground, single wall, excludes excavation &				
0090	Steel				
1000	500 gals	EA.	280	3,080	3,360
1020	1,000 gals	"	370	4,240	4,610
1980	Fiberglass, double wall				
2000	550 gals	EA.	370	8,800	9,170

13200.10	STORAGE TANKS, Cont'd...	UNIT	LABOR	MAT.	TOTAL
2020	1,000 gals	EA.	370	11,330	11,700
2520	Above ground				
2530	Steel, single wall				
2540	275 gals	EA.	220	1,760	1,980
2560	500 gals	"	370	4,400	4,770
2570	1,000 gals	"	450	6,050	6,500
3020	Fill cap	"	58.00	110	168
3040	Vent cap	"	58.00	110	168
3100	Level indicator	"	58.00	160	218

TABLE OF CONTENTS   PAGE

		UNIT	LABOR	MAT.	TOTAL
14210.10	**ELEVATORS**				
0120	Passenger elevators, electric, geared				
0502	Based on a shaft of 6 stops and 6 openings				
0510	50 fpm, 2000 lb	EA.	2,230	107,560	109,790
0520	100 fpm, 2000 lb	"	2,480	111,540	114,020
0525	150 fpm				
0530	2000 lb	EA.	2,790	123,490	126,280
0550	3000 lb	"	3,190	155,360	158,550
0560	4000 lb	"	3,720	161,340	165,060
1002	Based on a shaft of 8 stops and 8 openings				
1010	300 fpm				
1020	3000 lb	EA.	4,460	185,240	189,700
1040	3500 lb	"	4,460	189,380	193,840
1060	4000 lb	"	4,960	199,180	204,140
1080	5000 lb	"	5,310	223,090	228,400
1502	Hydraulic, based on a shaft of 3 stops, 3 openings				
1508	50 fpm				
1510	2000 lb	EA.	1,860	70,390	72,250
1520	2500 lb	"	1,860	75,330	77,190
1530	3000 lb	"	1,940	79,490	81,430
1600	For each additional; 50 fpm add per stop, $3500				
1620	500 lb, add per stop, $3500				
1630	Opening, add, $4200				
1640	Stop, add per stop, $5300				
1660	Bonderized steel door, add per opening, $400				
1670	Colored aluminum door, add per opening, $1500				
1680	Stainless steel door, add per opening, $650				
1690	Cast bronze door, add per opening, $1200				
1730	Custom cab interior, add per cab, $5000				
2000	Small elevators, 4 to 6 passenger capacity				
2005	Electric, push				
2010	2 stops	EA.	1,860	25,710	27,570
2020	3 stops	"	2,030	31,630	33,660
2030	4 stops	"	2,230	36,580	38,810
14300.10	**ESCALATORS**				
1000	Escalators				
1020	32" wide, floor to floor				
1040	12' high	EA.	3,720	140,070	143,790
1050	15' high	"	4,460	151,910	156,370
1060	18' high	"	5,580	163,750	169,330
1070	22' high	"	7,440	177,560	185,000
1080	25' high	"	8,920	197,280	206,200
1085	48" wide				
1090	12' high	EA.	3,850	155,850	159,700
1100	15' high	"	4,650	169,660	174,310
1120	18' high	"	5,870	181,500	187,370
1130	22' high	"	7,970	203,200	211,170
1140	25' high	"	8,920	216,910	225,830
14410.10	**PERSONNEL LIFTS**				
1000	Electrically operated, 1 or 2 person lift				
1001	With attached foot platforms				
1020	3 stops	EA.			9,770

		UNIT	LABOR	MAT.	TOTAL
14410.10	**PERSONNEL LIFTS, Cont'd...**				
1040	5 stops	EA.			15,510
1060	7 stops	"			17,930
2000	For each additional stop, add $1250				
3020	Residential stair climber, per story	EA.	440	4,590	5,030
14450.10	**VEHICLE LIFTS**				
1020	Automotive hoist, one post, semi-hydraulic, 8,000 lb	EA.	2,230	3,340	5,570
1040	Full hydraulic, 8,000 lb	"	2,230	3,410	5,640
1060	2 post, semi-hydraulic, 10,000 lb	"	3,190	3,460	6,650
1070	Full hydraulic				
1080	10,000 lb	EA.	3,190	4,270	7,460
1100	13,000 lb	"	5,580	5,180	10,760
1120	18,500 lb	"	5,580	8,030	13,610
1140	24,000 lb	"	5,580	10,590	16,170
1160	26,000 lb	"	5,580	11,360	16,940
1170	Pneumatic hoist, fully hydraulic				
1180	11,000 lb	EA.	7,440	5,760	13,200
1200	24,000 lb	"	7,440	10,100	17,540
14560.10	**CHUTES**				
1020	Linen chutes, stainless steel, with supports				
1030	18" dia.	L.F.	4.15	140	144
1040	24" dia.	"	4.47	170	174
1050	30" dia.	"	4.84	190	195
1060	Hopper	EA.	38.75	2,200	2,239
1070	Skylight	"	58.00	1,340	1,398
14580.10	**PNEUMATIC SYSTEMS**				
6000	Air-lift conveyor, 115 Volt, single phase, 100' long, 3" carrier	EA.	4,850	3,530	8,380
6010	4" carrier	"	4,850	4,380	9,230
6020	6" carrier	"	4,850	7,990	12,840
7000	Pneumatic tube system accessories				
7010	Couplings and hanging accessories for 3" carrier system	L.F.			5.91
7020	For 4" carrier system	"			8.42
7030	For 6" carrier system	"			11.25
7040	24" CL. Expanded 90° bend, heavy duty, 3"-6" carrier systems	EA.			100
7050	36" CL. Expanded 90° bend, heavy duty, 3"-6" carrier systems	"			250
7060	48" CL. Expanded 45° bend, heavy duty, 3" carrier system	"			80.00
7070	4" carrier system	"			110
7080	6" carrier system	"			170
7090	48" CL. Expanded 90° bend, heavy duty, 3" carrier system	"			100
8000	4" carrier system	"			140
8010	6" carrier system	"			250
8020	Stainless steel body up-grade, 3" carrier system	"			350
8030	4" carrier system	"			360
8040	6" carrier system	"			510
14580.30	**DUMBWAITERS**				
0010	28' travel, extruded alum., 4 stops, 100 lbs. capacity	EA.			5,400
0020	150 lbs. capacity	"			7,590
0030	200 lbs. capacity	"			10,560

Design & Construction Resources

TABLE OF CONTENTS   PAGE

		UNIT	LABOR	MAT.	TOTAL
15120.10	**BACKFLOW PREVENTERS**				
0080	Backflow preventer, flanged, cast iron, with valves				
0100	3" pipe	EA.	290	2,760	3,050
15410.06	**C.I. PIPE, BELOW GROUND**				
1010	No hub pipe				
1020	1-1/2" pipe	L.F.	2.88	6.18	9.06
1030	2" pipe	"	3.20	6.34	9.54
1120	3" pipe	"	3.60	8.75	12.35
1220	4" pipe	"	4.80	11.25	16.05
15410.10	**COPPER PIPE**				
0880	Type "K" copper				
0900	1/2"	L.F.	1.80	3.76	5.56
1000	3/4"	"	1.92	7.01	8.93
1020	1"	"	2.06	9.18	11.24
3000	DWV, copper				
3020	1-1/4"	L.F.	2.40	9.24	11.64
3030	1-1/2"	"	2.62	12.00	14.62
3040	2"	"	2.88	15.25	18.13
3070	3"	"	3.20	26.25	29.45
3080	4"	"	3.60	45.75	49.35
3090	6"	"	4.12	180	184
6080	Type "L" copper				
6090	1/4"	L.F.	1.69	1.51	3.20
6095	3/8"	"	1.69	2.33	4.02
6100	1/2"	"	1.80	2.70	4.50
6190	3/4"	"	1.92	4.32	6.24
6240	1"	"	2.06	6.50	8.56
6580	Type "M" copper				
6600	1/2"	L.F.	1.80	1.91	3.71
6620	3/4"	"	1.92	3.11	5.03
6630	1"	"	2.06	5.06	7.12
15410.11	**COPPER FITTINGS**				
0850	Slip coupling				
0860	1/4"	EA.	19.25	0.83	20.08
0870	1/2"	"	23.00	1.40	24.40
0880	3/4"	"	28.75	2.92	31.67
0890	1"	"	32.00	6.18	38.18
2660	Street ells, copper				
2670	1/4"	EA.	23.00	6.41	29.41
2680	3/8"	"	26.25	4.41	30.66
2690	1/2"	"	28.75	1.78	30.53
2700	3/4"	"	30.25	3.75	34.00
2710	1"	"	32.00	9.72	41.72
4190	DWV fittings, coupling with stop				
4210	1-1/2"	EA.	36.00	7.51	43.51
4230	2"	"	38.50	10.50	49.00
4260	3"	"	48.00	20.25	68.25
4280	3" x 2"	"	48.00	46.00	94.00
4290	4"	"	58.00	64.00	122
4300	Slip coupling				
4310	1-1/2"	EA.	36.00	11.75	47.75
4320	2"	"	38.50	14.00	52.50

		UNIT	LABOR	MAT.	TOTAL
15410.11	**COPPER FITTINGS, Cont'd...**				
4330	3"	EA.	48.00	25.25	73.25
4340	90 ells				
4350	1-1/2"	EA.	36.00	14.25	50.25
4360	1-1/2" x 1-1/4"	"	36.00	39.00	75.00
4370	2"	"	38.50	26.00	64.50
4380	2" x 1-1/2"	"	38.50	53.00	91.50
4390	3"	"	48.00	69.00	117
4400	4"	"	58.00	230	288
4410	Street, 90 elbows				
4420	1-1/2"	EA.	36.00	18.25	54.25
4430	2"	"	38.50	39.75	78.25
4440	3"	"	48.00	100	148
4450	4"	"	58.00	250	308
5410	No-hub adapters				
5420	1-1/2" x 2"	EA.	36.00	32.50	68.50
5430	2"	"	38.50	30.50	69.00
5440	2" x 3"	"	38.50	70.00	109
5450	3"	"	48.00	62.00	110
5460	3" x 4"	"	48.00	130	178
5470	4"	"	58.00	130	188
15410.82	**GALVANIZED STEEL PIPE**				
1000	Galvanized pipe				
1020	1/2" pipe	L.F.	5.76	2.78	8.54
1040	3/4" pipe	"	7.21	3.63	10.84
1200	90 degree ell, 150 lb malleable iron, galvanized				
1210	1/2"	EA.	11.50	2.15	13.65
1220	3/4"	"	14.50	2.86	17.36
1400	45 degree ell, 150 lb m.i., galv.				
1410	1/2"	EA.	11.50	3.43	14.93
1420	3/4"	"	14.50	4.66	19.16
1520	Tees, straight, 150 lb m.i., galv.				
1530	1/2"	EA.	14.50	2.86	17.36
1540	3/4"	"	16.50	4.75	21.25
1800	Couplings, straight, 150 lb m.i., galv.				
1810	1/2"	EA.	11.50	2.64	14.14
1820	3/4"	"	12.75	3.16	15.91
15430.23	**CLEANOUTS**				
0980	Cleanout, wall				
1000	2"	EA.	38.50	160	199
1020	3"	"	38.50	230	269
1040	4"	"	48.00	240	288
1050	Floor				
1060	2"	EA.	48.00	150	198
1080	3"	"	48.00	200	248
1100	4"	"	58.00	200	258
15430.25	**HOSE BIBBS**				
0005	Hose bibb				
0010	1/2"	EA.	19.25	9.07	28.32
0200	3/4"	"	19.25	9.62	28.87

		UNIT	LABOR	MAT.	TOTAL
15430.60	**VALVES**				
0780	Gate valve, 125 lb, bronze, soldered				
0800	1/2"	EA.	14.50	25.75	40.25
1000	3/4"	"	14.50	31.00	45.50
1280	Check valve, bronze, soldered, 125 lb				
1300	1/2"	EA.	14.50	40.50	55.00
1320	3/4"	"	14.50	50.00	64.50
1790	Globe valve, bronze, soldered, 125 lb				
1800	1/2"	EA.	16.50	60.00	76.50
1810	3/4"	"	18.00	74.00	92.00
15430.65	**VACUUM BREAKERS**				
1000	Vacuum breaker, atmospheric, threaded connection				
1010	3/4"	EA.	23.00	45.75	68.75
1018	Anti-siphon, brass				
1020	3/4"	EA.	23.00	49.50	72.50
15430.68	**STRAINERS**				
0980	Strainer, Y pattern, 125 psi, cast iron body, threaded				
1000	3/4"	EA.	20.50	11.50	32.00
1980	250 psi, brass body, threaded				
2000	3/4"	EA.	23.00	29.75	52.75
2130	Cast iron body, threaded				
2140	3/4"	EA.	23.00	17.50	40.50
15430.70	**DRAINS, ROOF & FLOOR**				
1020	Floor drain, cast iron, with cast iron top				
1030	2"	EA.	48.00	130	178
1040	3"	"	48.00	140	188
1050	4"	"	48.00	290	338
1090	Roof drain, cast iron				
1100	2"	EA.	48.00	210	258
1110	3"	"	48.00	220	268
1120	4"	"	48.00	280	328
15440.10	**BATHS**				
0980	Bath tub, 5' long				
1000	Minimum	EA.	190	530	720
1020	Average	"	290	1,160	1,450
1040	Maximum	"	580	2,640	3,220
1050	6' long				
1060	Minimum	EA.	190	590	780
1080	Average	"	290	1,210	1,500
1100	Maximum	"	580	3,420	4,000
1110	Square tub, whirlpool, 4'x4'				
1120	Minimum	EA.	290	1,810	2,100
1140	Average	"	580	2,570	3,150
1160	Maximum	"	720	7,850	8,570
1170	5'x5'				
1180	Minimum	EA.	290	1,810	2,100
1200	Average	"	580	2,570	3,150
1220	Maximum	"	720	8,000	8,720
1230	6'x6'				
1240	Minimum	EA.	290	2,210	2,500
1260	Average	"	580	3,230	3,810
1280	Maximum	"	720	9,270	9,990

		UNIT	LABOR	MAT.	TOTAL
15440.10	**BATHS, Cont'd...**				
8980	For trim and rough-in				
9000	Minimum	EA.	190	190	380
9020	Average	"	290	280	570
9040	Maximum	"	580	780	1,360
15440.12	**DISPOSALS & ACCESSORIES**				
0040	Continuous feed				
0050	Minimum	EA.	120	72.00	192
0060	Average	"	140	200	340
0070	Maximum	"	190	390	580
0200	Batch feed, 1/2 hp				
0220	Minimum	EA.	120	280	400
0230	Average	"	140	550	690
0240	Maximum	"	190	950	1,140
1100	Hot water dispenser				
1110	Minimum	EA.	120	200	320
1120	Average	"	140	320	460
1130	Maximum	"	190	510	700
1140	Epoxy finish faucet	"	120	290	410
1160	Lock stop assembly	"	72.00	61.00	133
1170	Mounting gasket	"	48.00	7.04	55.04
1180	Tailpipe gasket	"	48.00	1.03	49.03
1190	Stopper assembly	"	58.00	24.00	82.00
1200	Switch assembly, on/off	"	96.00	27.50	124
1210	Tailpipe gasket washer	"	28.75	1.10	29.85
1220	Stop gasket	"	32.00	2.42	34.42
1230	Tailpipe flange	"	28.75	0.27	29.02
1240	Tailpipe	"	36.00	3.13	39.13
15440.15	**FAUCETS**				
0980	Kitchen				
1000	Minimum	EA.	96.00	83.00	179
1020	Average	"	120	230	350
1040	Maximum	"	140	290	430
1050	Bath				
1060	Minimum	EA.	96.00	83.00	179
1080	Average	"	120	240	360
1100	Maximum	"	140	370	510
1110	Lavatory, domestic				
1120	Minimum	EA.	96.00	88.00	184
1140	Average	"	120	280	400
1160	Maximum	"	140	460	600
1290	Washroom				
1300	Minimum	EA.	96.00	110	206
1320	Average	"	120	280	400
1340	Maximum	"	140	510	650
1350	Handicapped				
1360	Minimum	EA.	120	120	240
1380	Average	"	140	360	500
1400	Maximum	"	190	560	750
1410	Shower				
1420	Minimum	EA.	96.00	110	206
1440	Average	"	120	320	440

		UNIT	LABOR	MAT.	TOTAL
15440.15	**FAUCETS, Cont'd...**				
1460	Maximum	EA.	140	510	650
1480	For trim and rough-in				
1500	Minimum	EA.	120	77.00	197
1520	Average	"	140	120	260
1540	Maximum	"	290	200	490
15440.18	**HYDRANTS**				
0980	Wall hydrant				
1000	8" thick	EA.	96.00	360	456
1020	12" thick	"	120	430	550
15440.20	**LAVATORIES**				
1980	Lavatory, counter top, porcelain enamel on cast iron				
2000	Minimum	EA.	120	190	310
2010	Average	"	140	290	430
2020	Maximum	"	190	520	710
2080	Wall hung, china				
2100	Minimum	EA.	120	260	380
2110	Average	"	140	310	450
2120	Maximum	"	190	770	960
2280	Handicapped				
2300	Minimum	EA.	140	430	570
2310	Average	"	190	500	690
2320	Maximum	"	290	830	1,120
8980	For trim and rough-in				
9000	Minimum	EA.	140	220	360
9020	Average	"	190	370	560
9040	Maximum	"	290	460	750
15440.30	**SHOWERS**				
0980	Shower, fiberglass, 36"x34"x84"				
1000	Minimum	EA.	410	570	980
1020	Average	"	580	800	1,380
1040	Maximum	"	580	1,160	1,740
2980	Steel, 1 piece, 36"x36"				
3000	Minimum	EA.	410	530	940
3020	Average	"	580	800	1,380
3040	Maximum	"	580	950	1,530
3980	Receptor, molded stone, 36"x36"				
4000	Minimum	EA.	190	220	410
4020	Average	"	290	370	660
4040	Maximum	"	480	570	1,050
8980	For trim and rough-in				
9000	Minimum	EA.	260	220	480
9020	Average	"	320	370	690
9040	Maximum	"	580	460	1,040
15440.40	**SINKS**				
0980	Service sink, 24"x29"				
1000	Minimum	EA.	140	640	780
1020	Average	"	190	790	980
1040	Maximum	"	290	1,170	1,460
2000	Kitchen sink, single, stainless steel, single bowl				
2020	Minimum	EA.	120	280	400
2040	Average	"	140	320	460

		UNIT	LABOR	MAT.	TOTAL
15440.40	**SINKS, Cont'd...**				
2060	Maximum	EA.	190	580	770
2070	Double bowl				
2080	Minimum	EA.	140	320	460
2100	Average	"	190	360	550
2120	Maximum	"	290	620	910
2190	Porcelain enamel, cast iron, single bowl				
2200	Minimum	EA.	120	200	320
2220	Average	"	140	260	400
2240	Maximum	"	190	410	600
2250	Double bowl				
2260	Minimum	EA.	140	280	420
2280	Average	"	190	390	580
2300	Maximum	"	290	550	840
2980	Mop sink, 24"x36"x10"				
3000	Minimum	EA.	120	480	600
3020	Average	"	140	580	720
3040	Maximum	"	190	780	970
5980	Washing machine box				
6000	Minimum	EA.	140	180	320
6040	Average	"	190	250	440
6060	Maximum	"	290	310	600
8980	For trim and rough-in				
9000	Minimum	EA.	190	290	480
9020	Average	"	290	440	730
9040	Maximum	"	380	560	940
15440.60	**WATER CLOSETS**				
0980	Water closet flush tank, floor mounted				
1000	Minimum	EA.	140	330	470
1010	Average	"	190	650	840
1020	Maximum	"	290	1,020	1,310
1030	Handicapped				
1040	Minimum	EA.	190	370	560
1050	Average	"	290	670	960
1060	Maximum	"	580	1,280	1,860
8980	For trim and rough-in				
9000	Minimum	EA.	140	210	350
9020	Average	"	190	250	440
9040	Maximum	"	290	330	620
15440.70	**WATER HEATERS**				
0980	Water heater, electric				
1000	6 gal	EA.	96.00	350	446
1020	10 gal	"	96.00	360	456
1030	15 gal	"	96.00	360	456
1040	20 gal	"	120	500	620
1050	30 gal	"	120	520	640
1060	40 gal	"	120	560	680
1070	52 gal	"	140	630	770
2980	Oil fired				
3000	20 gal	EA.	290	1,300	1,590
3020	50 gal	"	410	2,020	2,430

		UNIT	LABOR	MAT.	TOTAL
15610.10	**FURNACES**				
0980	Electric, hot air				
1000	40 mbh	EA.	290	810	1,100
1020	60 mbh	"	300	880	1,180
1040	80 mbh	"	320	960	1,280
1060	100 mbh	"	340	1,080	1,420
1080	125 mbh	"	350	1,320	1,670
1980	Gas fired hot air				
2000	40 mbh	EA.	290	810	1,100
2020	60 mbh	"	300	870	1,170
2040	80 mbh	"	320	1,000	1,320
2060	100 mbh	"	340	1,040	1,380
2080	125 mbh	"	350	1,140	1,490
2980	Oil fired hot air				
3000	40 mbh	EA.	290	1,090	1,380
3020	60 mbh	"	300	1,800	2,100
3040	80 mbh	"	320	1,820	2,140
3060	100 mbh	"	340	1,850	2,190
3080	125 mbh	"	350	1,910	2,260
15780.20	**ROOFTOP UNITS**				
0980	Packaged, single zone rooftop unit, with roof curb				
1000	2 ton	EA.	580	3,400	3,980
1020	3 ton	"	580	3,580	4,160
1040	4 ton	"	720	3,900	4,620
15830.70	**UNIT HEATERS**				
0980	Steam unit heater, horizontal				
1000	12,500 btuh, 200 cfm	EA.	96.00	500	596
1010	17,000 btuh, 300 cfm	"	96.00	660	756
15855.10	**AIR HANDLING UNITS**				
0980	Air handling unit, medium pressure, single zone				
1000	1500 cfm	EA.	360	4,000	4,360
1060	3000 cfm	"	640	5,260	5,900
8980	Rooftop air handling units				
9000	4950 cfm	EA.	640	11,500	12,140
9060	7370 cfm	"	820	14,580	15,400
15870.20	**EXHAUST FANS**				
0160	Belt drive roof exhaust fans				
1020	640 cfm, 2618 fpm	EA.	72.00	1,030	1,102
1030	940 cfm, 2604 fpm	"	72.00	1,340	1,412
15890.10	**METAL DUCTWORK**				
0090	Rectangular duct				
0100	Galvanized steel				
1000	Minimum	Lb.	5.24	0.88	6.12
1010	Average	"	6.40	1.10	7.50
1020	Maximum	"	9.61	1.68	11.29
1080	Aluminum				
1100	Minimum	Lb.	11.50	2.29	13.79
1120	Average	"	14.50	3.05	17.55
1140	Maximum	"	19.25	3.79	23.04
1160	Fittings				
1180	Minimum	EA.	19.25	7.26	26.51
1200	Average	"	28.75	11.00	39.75

DIVISION # 15 MECHANICAL

		UNIT	LABOR	MAT.	TOTAL
15890.10	**METAL DUCTWORK, Cont'd...**				
1220	Maximum	EA.	58.00	16.00	74.00
15890.30	**FLEXIBLE DUCTWORK**				
1010	Flexible duct, 1.25" fiberglass				
1020	5" dia.	L.F.	2.88	3.31	6.19
1040	6" dia.	"	3.20	3.68	6.88
1060	7" dia.	"	3.39	4.54	7.93
1080	8" dia.	"	3.60	4.76	8.36
1100	10" dia.	"	4.12	6.34	10.46
1120	12" dia.	"	4.43	6.93	11.36
9000	Flexible duct connector, 3" wide fabric	"	9.61	2.31	11.92
15910.10	**DAMPERS**				
0980	Horizontal parallel aluminum backdraft damper				
1000	12" x 12"	EA.	14.50	55.00	69.50
1010	16" x 16"	"	16.50	57.00	73.50
15940.10	**DIFFUSERS**				
1980	Ceiling diffusers, round, baked enamel finish				
2000	6" dia.	EA.	19.25	45.75	65.00
2020	8" dia.	"	24.00	55.00	79.00
2040	10" dia.	"	24.00	61.00	85.00
2060	12" dia.	"	24.00	78.00	102
2480	Rectangular				
2500	6x6"	EA.	19.25	48.75	68.00
2520	9x9"	"	28.75	59.00	87.75
2540	12x12"	"	28.75	86.00	115
2560	15x15"	"	28.75	110	139
2580	18x18"	"	28.75	140	169

Design & Construction Resources

TABLE OF CONTENTS   PAGE

		UNIT	LABOR	MAT.	TOTAL
16050.30	**BUS DUCT**				
1000	Bus duct, 100a, plug-in				
1010	10', 600v	EA.	180	230	410
1020	With ground	"	280	310	590
1145	Circuit breakers, with enclosure				
1147	1 pole				
1150	15a-60a	EA.	67.00	230	297
1160	70a-100a	"	83.00	260	343
1165	2 pole				
1170	15a-60a	EA.	73.00	340	413
1180	70a-100a	"	87.00	410	497
16110.22	**EMT CONDUIT**				
0080	EMT conduit				
0100	1/2"	L.F.	2.01	0.56	2.57
1020	3/4"	"	2.66	1.02	3.68
1030	1"	"	3.33	1.70	5.03
2980	90 deg. elbow				
3000	1/2"	EA.	5.93	5.26	11.19
3040	3/4"	"	6.67	5.78	12.45
3060	1"	"	7.11	8.92	16.03
3980	Connector, steel compression				
4000	1/2"	EA.	5.93	1.41	7.34
4040	3/4"	"	5.93	2.69	8.62
4060	1"	"	5.93	4.06	9.99
0080	Flexible conduit, steel				
0100	3/8"	L.F.	2.01	0.52	2.53
1020	1/2	"	2.01	0.59	2.60
1040	3/4"	"	2.66	0.81	3.47
1060	1"	"	2.66	1.53	4.19
16110.24	**GALVANIZED CONDUIT**				
1980	Galvanized rigid steel conduit				
2000	1/2"	L.F.	2.66	2.48	5.14
2040	3/4"	"	3.33	2.74	6.07
2060	1"	"	3.95	3.96	7.91
2080	1-1/4"	"	5.33	5.48	10.81
2100	1-1/2"	"	5.93	6.44	12.37
2120	2"	"	6.67	8.20	14.87
2480	90 degree ell				
2500	1/2"	EA.	16.75	10.25	27.00
2540	3/4"	"	20.50	10.75	31.25
2560	1"	"	25.50	16.50	42.00
2580	1-1/4"	"	29.75	22.75	52.50
2590	1-1/2"	"	33.25	28.00	61.25
2600	2"	"	35.50	40.75	76.25
3200	Couplings, with set screws				
3220	1/2"	EA.	3.33	5.17	8.50
3260	3/4"	"	3.95	6.82	10.77
3280	1"	"	5.33	11.00	16.33
3300	1-1/4"	"	6.67	18.50	25.17
3320	1-1/2"	"	8.21	24.00	32.21
3340	2"	"	9.70	54.00	63.70

		UNIT	LABOR	MAT.	TOTAL
16110.25	**PLASTIC CONDUIT**				
3010	PVC conduit, schedule 40				
3020	1/2"	L.F.	2.01	0.63	2.64
3040	3/4"	"	2.01	0.79	2.80
3060	1"	"	2.66	1.14	3.80
3080	1-1/4"	"	2.66	1.58	4.24
3100	1-1/2"	"	3.33	1.88	5.21
3120	2"	"	3.33	2.40	5.73
3480	Couplings				
3500	1/2"	EA.	3.33	0.42	3.75
3520	3/4"	"	3.33	0.51	3.84
3540	1"	"	3.33	0.80	4.13
3560	1-1/4"	"	3.95	1.05	5.00
3580	1-1/2"	"	3.95	1.46	5.41
3600	2"	"	3.95	1.92	5.87
3705	90 degree elbows				
3710	1/2"	EA.	6.67	1.66	8.33
3740	3/4"	"	8.21	1.81	10.02
3760	1"	"	8.21	2.87	11.08
3780	1-1/4"	"	9.70	4.00	13.70
3800	1-1/2"	"	12.75	5.42	18.17
3810	2"	"	14.75	7.56	22.31
16110.28	**STEEL CONDUIT**				
7980	Intermediate metal conduit (IMC)				
8000	1/2"	L.F.	2.01	1.54	3.55
8040	3/4"	"	2.66	1.89	4.55
8060	1"	"	3.33	2.86	6.19
8080	1-1/4"	"	3.95	3.66	7.61
8100	1-1/2"	"	5.33	4.57	9.90
8120	2"	"	5.93	5.97	11.90
8490	90 degree ell				
8500	1/2"	EA.	16.75	12.75	29.50
8540	3/4"	"	20.50	13.50	34.00
8560	1"	"	25.50	20.50	46.00
8580	1-1/4"	"	29.75	28.50	58.25
8600	1-1/2"	"	33.25	35.25	68.50
8620	2"	"	38.25	51.00	89.25
9260	Couplings				
9280	1/2"	EA.	3.33	3.14	6.47
9290	3/4"	"	3.95	3.86	7.81
9300	1"	"	5.33	5.72	11.05
9310	1-1/4"	"	5.93	7.16	13.09
9320	1-1/2"	"	6.67	9.05	15.72
9330	2"	"	7.11	12.00	19.11
16110.35	**SURFACE MOUNTED RACEWAY**				
0980	Single Raceway				
1000	3/4" x 17/32" Conduit	L.F.	2.66	1.67	4.33
1020	Mounting Strap	EA.	3.55	0.45	4.00
1040	Connector	"	3.55	0.60	4.15
1060	Elbow				
2000	45 degree	EA.	3.33	7.62	10.95
2020	90 degree	"	3.33	2.43	5.76

		UNIT	LABOR	MAT.	TOTAL
16110.35	**SURFACE MOUNTED RACEWAY, Cont'd...**				
2040	internal	EA.	3.33	3.05	6.38
2050	external	"	3.33	2.82	6.15
2060	Switch	"	26.75	19.75	46.50
2100	Utility Box	"	26.75	13.25	40.00
2110	Receptacle	"	26.75	23.50	50.25
2140	3/4" x 21/32" Conduit	L.F.	2.66	1.90	4.56
2160	Mounting Strap	EA.	3.55	0.70	4.25
2180	Connector	"	3.55	0.72	4.27
2200	Elbow				
2210	45 degree	EA.	3.33	9.41	12.74
2220	90 degree	"	3.33	2.59	5.92
2240	internal	"	3.33	3.52	6.85
2260	external	"	3.33	3.52	6.85
3000	Switch	"	26.75	19.75	46.50
3010	Utility Box	"	26.75	13.25	40.00
3020	Receptacle	"	26.75	23.50	50.25
16120.43	**COPPER CONDUCTORS**				
0980	Copper conductors, type THW, solid				
1000	#14	L.F.	0.26	0.12	0.38
1040	#12	"	0.33	0.18	0.51
1060	#10	"	0.40	0.28	0.68
2010	THHN-THWN, solid				
2020	#14	L.F.	0.26	0.12	0.38
2040	#12	"	0.33	0.18	0.51
2060	#10	"	0.40	0.28	0.68
6215	Type "BX" solid armored cable				
6220	#14/2	L.F.	1.66	0.82	2.48
6230	#14/3	"	1.87	1.29	3.16
6240	#14/4	"	2.05	1.81	3.86
6250	#12/2	"	1.87	0.84	2.71
6260	#12/3	"	2.05	1.35	3.40
6270	#12/4	"	2.32	1.87	4.19
6280	#10/2	"	2.05	1.56	3.61
6290	#10/3	"	2.32	2.23	4.55
6300	#10/4	"	2.66	3.47	6.13
16120.47	**SHEATHED CABLE**				
6700	Non-metallic sheathed cable				
6705	Type NM cable with ground				
6710	#14/2	L.F.	0.99	0.35	1.34
6720	#12/2	"	1.06	0.53	1.59
6730	#10/2	"	1.18	0.85	2.03
6740	#8/2	"	1.33	1.39	2.72
6750	#6/2	"	1.66	2.20	3.86
6760	#14/3	"	1.72	0.49	2.21
6770	#12/3	"	1.77	0.77	2.54
6780	#10/3	"	1.80	1.22	3.02
6790	#8/3	"	1.84	2.05	3.89
6800	#6/3	"	1.87	3.32	5.19
6810	#4/3	"	2.13	6.87	9.00
6820	#2/3	"	2.32	10.25	12.57

		UNIT	LABOR	MAT.	TOTAL
16130.40	**BOXES**				
5000	Round cast box, type SEH				
5010	1/2"	EA.	23.25	20.00	43.25
5020	3/4"	"	28.00	20.00	48.00
16130.60	**PULL AND JUNCTION BOXES**				
1050	4"				
1060	Octagon box	EA.	7.62	3.33	10.95
1070	Box extension	"	3.95	5.61	9.56
1080	Plaster ring	"	3.95	3.08	7.03
1100	Cover blank	"	3.95	1.36	5.31
1120	Square box	"	7.62	4.79	12.41
1140	Box extension	"	3.95	4.69	8.64
1160	Plaster ring	"	3.95	2.57	6.52
1180	Cover blank	"	3.95	1.32	5.27
16130.80	**RECEPTACLES**				
0500	Contractor grade duplex receptacles, 15a 120v				
0510	Duplex	EA.	13.25	1.46	14.71
1000	125 volt, 20a, duplex, grounding type, standard grade	"	13.25	10.75	24.00
1040	Ground fault interrupter type	"	19.75	35.25	55.00
1520	250 volt, 20a, 2 pole, single receptacle, ground type	"	13.25	18.25	31.50
1540	120/208v, 4 pole, single receptacle, twist lock				
1560	20a	EA.	23.25	21.50	44.75
1580	50a	"	23.25	41.00	64.25
1590	125/250v, 3 pole, flush receptacle				
1600	30a	EA.	19.75	21.75	41.50
1620	50a	"	19.75	27.00	46.75
1640	60a	"	23.25	70.00	93.25
16350.10	**CIRCUIT BREAKERS**				
5000	Load center circuit breakers, 240v				
5010	1 pole, 10-60a	EA.	16.75	16.50	33.25
5015	2 pole				
5020	10-60a	EA.	26.75	38.50	65.25
5030	70-100a	"	44.50	120	165
5040	110-150a	"	48.50	250	299
5065	Load center, G.F.I. breakers, 240v				
5070	1 pole, 15-30a	EA.	19.75	140	160
5095	Tandem breakers, 240v				
5100	1 pole, 15-30a	EA.	26.75	31.25	58.00
5110	2 pole, 15-30a	"	35.50	57.00	92.50
16395.10	**GROUNDING**				
0500	Ground rods, copper clad, 1/2" x				
0510	6'	EA.	44.50	13.25	57.75
0520	8'	"	48.50	18.50	67.00
0535	5/8" x				
0550	6'	EA.	48.50	17.75	66.25
0560	8'	"	67.00	23.00	90.00
16430.20	**METERING**				
0500	Outdoor wp meter sockets, 1 gang, 240v, 1 phase				
0510	Includes sealing ring, 100a	EA.	100	47.50	148
0520	150a	"	120	63.00	183
0530	200a	"	130	80.00	210

		UNIT	LABOR	MAT.	TOTAL
16470.10	**PANELBOARDS**				
1000	Indoor load center, 1 phase 240v main lug only				
1020	30a - 2 spaces	EA.	130	24.25	154
1030	100a - 8 spaces	"	160	77.00	237
1040	150a - 16 spaces	"	200	200	400
1050	200a - 24 spaces	"	230	420	650
1060	200a - 42 spaces	"	270	430	700
16490.10	**SWITCHES**				
4000	Photo electric switches				
4010	1000 watt				
4020	105-135v	EA.	48.50	33.50	82.00
4970	Dimmer switch and switch plate				
4990	600w	EA.	20.50	30.75	51.25
5171	Contractor grade wall switch 15a, 120v				
5172	Single pole	EA.	10.75	1.62	12.37
5173	Three way	"	13.25	2.97	16.22
5174	Four way	"	17.75	10.00	27.75
16510.05	**INTERIOR LIGHTING**				
0005	Recessed fluorescent fixtures, 2'x2'				
0010	2 lamp	EA.	48.50	63.00	112
0020	4 lamp	"	48.50	85.00	134
0205	Surface mounted incandescent fixtures				
0210	40w	EA.	44.50	95.00	140
0220	75w	"	44.50	98.00	143
0230	100w	"	44.50	110	155
0240	150w	"	44.50	140	185
0289	Recessed incandescent fixtures				
0290	40w	EA.	100	130	230
0300	75w	"	100	140	240
0310	100w	"	100	150	250
0320	150w	"	100	160	260
0395	Light track single circuit				
0400	2'	EA.	33.25	38.50	71.75
0410	4'	"	33.25	45.50	78.75
0420	8'	"	67.00	62.00	129
0430	12'	"	100	88.00	188
16510.10	**LIGHTING INDUSTRIAL**				
0500	Strip fluorescent				
0510	4'				
0520	1 lamp	EA.	44.50	43.25	87.75
0540	2 lamps	"	44.50	53.00	97.50
0550	8'				
0560	1 lamp	EA.	48.50	63.00	112
0580	2 lamps	"	59.00	95.00	154
1000	Parabolic troffer, 2'x2'				
1020	With 2 "U" lamps	EA.	67.00	120	187
1060	With 3 "U" lamps	"	76.00	140	216
1080	2'x4'				
1100	With 2 40w lamps	EA.	76.00	140	216
1120	With 3 40w lamps	"	89.00	140	229
1140	With 4 40w lamps	"	89.00	150	239
3120	High pressure sodium, hi-bay open				

		UNIT	LABOR	MAT.	TOTAL
16510.10	**LIGHTING INDUSTRIAL, Cont'd...**				
3140	400w	EA.	120	430	550
3160	1000w	"	160	750	910
3170	Enclosed				
3180	400w	EA.	160	700	860
3200	1000w	"	200	980	1,180
3210	Metal halide hi-bay, open				
3220	400w	EA.	120	270	390
3240	1000w	"	160	550	710
3250	Enclosed				
3260	400w	EA.	160	610	770
3280	1000w	"	200	580	780
3590	Metal halide, low bay, pendant mounted				
3600	175w	EA.	89.00	350	439
3620	250w	"	110	480	590
3660	400w	"	150	520	670
16710.10	**COMMUNICATIONS COMPONENTS**				
0020	Port desktop switch unit-4	EA.			77.00
0030	(USB)-4	"			200
0040	8	"			330
0050	16	"			550
0060	32	"			3,520
0070	Port console unit-8	"			2,200
0080	16	"			2,750
0090	Cat 5EJack and RJ45 coupler	"			3.74
1000	Quick-Port and voice grade	"			4.23
1010	Trac-Jack category 5E and connectors	"			4.21
1020	Snap-In connector	"			5.77
1030	Fast ethernet media converter	"			140
1040	Gigabit media converter	"			390
1050	With link fault signaling	"			130
1060	Gigabit switching media converter	"			550
1070	16-Bay media chassis	"			670
1080	Fiber-optic cable, Single-Mode, 50/125 microns, 10' length	"			22.00
1090	Multi-Mode, 62.5/125 microns, 10' length	"			25.25
2000	Fiber-Optic connectors, Unicam	"			17.00
2010	Fast-cam	"			17.50
2020	Adhesive style	"			6.32
2030	Threat-lock	"			10.75
2040	Unicam, high performance, single-mode	"			22.00
2050	multi-mode	"			19.75
2060	Network Cable, Cat5, solid PVC, 50'	"			16.00
2070	1000' Cat5, stranded PVC	"			190
2080	plenum PVC	"			220
2090	1000' Cat6, solid PVC	"			150
3000	USB Cables, 5 in 1 connector, male and female	"			19.75
3010	3 in 1 quick-connect, 4-pin or 6-pin	"			24.25
3020	Squid hub	"			33.00
16760.10	**AUDIO/VIDEO COMPONENTS**				
0010	Key-stone jack module	EA.			3.30
0020	F-Type quick port	"			2.31
0030	BNC quick port	"			5.50

		UNIT	LABOR	MAT.	TOTAL
16760.10	**AUDIO/VIDEO COMPONENTS, Cont'd...**				
0040	RCA jack bulkhead connector	EA.			4.40
0050	Quick-Port	"			5.72
0060	Snap-in	"			8.80
0070	HDMI wall plate and jack	"			10.50
0080	Voice/Data adapter	"			3.85
16760.20	**AUDIO/VIDEO CABLES**				
0010	Monster cable, 24k. gold plated	EA.			27.50
0020	DVI-HDMI	"			11.75
0030	Satellite/Video	"			3.96
0040	Digital/Optical	"			12.75
0050	Video/RCA	"			7.42
0060	S-Video	"			8.25
0070	Monster/S-video	"			44.00
0080	Speaker, clear jacket	FT.			0.93
0090	Speaker, high performance	"			2.20
0100	Flex-Premiere, oxygen free	"			0.33
16850.10	**ELECTRIC HEATING**				
1000	Baseboard heater				
1020	2', 375w	EA.	67.00	41.75	109
1040	3', 500w	"	67.00	49.50	117
1060	4', 750w	"	76.00	55.00	131
1100	5', 935w	"	89.00	78.00	167
1120	6', 1125w	"	110	92.00	202
1140	7', 1310w	"	120	100	220
1160	8', 1500w	"	130	120	250
1180	9', 1680w	"	150	130	280
1200	10', 1875w	"	150	180	330
1210	Unit heater, wall mounted				
1215	750w	EA.	110	160	270
1220	1500w	"	110	220	330
16910.40	**CONTROL CABLE**				
0980	Control cable, 600v, #14 THWN, PVC jacket				
1000	2 wire	L.F.	0.53	0.31	0.84
1020	4 wire	"	0.66	0.53	1.19

Average Square Foot Costs
2011

Housing (with Wood Frame)

Square Foot Costs do not include Garages, Land, Furnishings, Equipment, Landscaping, Financing, or Architect's Fee. Total Costs, except General Contractor's Fee, are included in each division. See Page 2 for Metropolitan Cost Variation Modifier. See Division 13 for Metal Buildings, Greenhouses, and Air-Supported Structures.

	Division	Houses - 3 Bedrooms with Basement			Houses - 3 Bedrooms without Basement		
		Low Rent	Tract or Project	Custom	Low Rent	Tract or Project	Custom
1.	General Conditions	$5.07	$5.18	$5.62	$6.57	$6.79	$7.24
2.	Site Work	4.01	4.58	4.93	6.76	7.08	7.34
3.	Concrete	4.64	4.87	5.18	8.41	8.62	9.56
4.	Masonry	5.68	5.90	7.85	8.07	8.42	11.16
5.	Steel	1.84	1.84	1.98	.91	.89	.92
6.	Carpentry	18.87	18.52	20.87	27.66	27.43	29.84
7.	Moisture & Therm. Prot.	4.82	4.82	4.95	6.72	6.96	7.56
8.	Doors, Windows & Glass	9.76	9.42	9.82	12.03	12.37	12.97
9.	Finishes	8.90	9.71	11.85	13.18	13.39	18.96
10.	Specialties	.90	.92	1.01	.99	1.00	1.33
11.	Equipment (Cabinets)	3.29	3.35	5.08	5.90	5.85	8.81
12.	Furnishings	.67	.67	0.73	1.03	1.03	1.03
13.	Special Construction	-	-	-	-	-	-
14.	Conveying	-	-	-	-	-	-
15.	Mechanical: Plumbing	7.87	7.93	9.33	11.65	11.59	14.41
	Heating	4.77	4.94	5.14	6.71	7.08	10.00
	Air Cond.	-	-	-	-	-	-
16.	Electrical	4.59	4.59	5.53	6.82	6.82	8.10
	SUB TOTAL	$85.69	$87.26	$99.87	$123.44	$125.34	$149.23
	CONTRACTOR'S SERVICES	7.45	7.65	8.72	10.80	11.11	14.51
	TOTAL AVERAGE Sq.Ft. Cost	**$93.14**	**$94.91**	**$108.59**	**$134.24**	**$136.46**	**$163.74**

Unfinished Basement, Storage and Utility Areas included in Sq.Ft. Costs:				No Basement Utilities In Finished Area		
Average Cost/ Unit	$186,276	$189,819	$271,471	$161,088	$163,748	$245,614
Average Sq.Ft. Size/Unit (includes basement)	2,000	2,000	2,500	1,200	1,200	1,500
Add Architect's Fee	$ 7,667	$8,587	$19,628	$6,624	$7,483	$14,721

	COST	
	Sq. Ft.	Ea. Unit
Add for Finishing Basement Space For Bedrooms or Amusement Rooms, 12' x 20'	$35.43	$8,504
Add for Single Garage (no Interior Finish), 12' x 20'	$33.74	$8,098
Add for Double Garage (no Interior Finish), 20 x 20'	$35.43	$8,504
Add (or Deduct) for Bedrooms, 12' x 12'	$50.39	$7,256
Add per Bathroom with 2 Fixtures, 6' x 8'	$157.92	$7,580
Add per Bathroom with 3 Fixtures, 8' x 8'	$157.92	$10,107
Add for Fireplace	-	$7,873
Add for Central Air Conditioning	-	$6,524
Add for Drain Tiling	-	$3,555
Add for Well	-	$8,098
Add for Septic System	-	$8,098

Average Square Foot Costs
2011

Apartments, Hotels

Square Foot Costs do not include Land, Furnishings, Equipment, Landscaping and Financing.

		Apartments			Motels & Hotels		
	Division	2 & 3 Story	2 & 3 Story	High Rise	High Rise	1 & 2 Story	High Rise
		(1)a	(1)b	(2)	(3)	(1)b	(2)
1.	General Conditions	$ 4.79	$ 5.07	$7.24	$ 7.74	$ 5.24	$ 6.68
2.	Site Work	3.45	3.39	4.89	4.81	7.11	5.50
3.	Concrete	2.73	2.72	26.01	17.56	3.19	26.64
4.	Masonry	1.95	12.36	3.21	3.06	12.31	15.92
5.	Steel	1.74	1.75	2.78	2.90	2.11	2.66
6.	Carpentry and Millwork	19.80	16.61	6.41	5.94	16.63	5.16
7.	Moisture Protection	4.43	4.68	2.38	2.34	4.31	2.38
8.	Doors, Windows and Glass	6.92	7.54	8.57	8.32	8.64	10.31
9.	Finishes	13.56	13.60	19.93	13.12	13.56	18.68
10.	Specialties	1.01	1.02	1.06	1.05	1.24	1.26
11.	Equipment	2.09	2.76	2.13	2.09	3.46	3.69
12.	Furnishings	1.03	1.12	1.19	1.21	2.42	1.84
13.	Special Construction	-	-	-	-	-	-
14.	Conveying	-	2.90	4.37	4.51	-	5.07
15.	Mechanical:						
	Plumbing	9.33	9.33	9.15	8.99	10.60	14.29
	Heating and Ventilation	7.15	7.73	8.56	8.22	7.43	9.65
	Air Conditioning	2.55	2.55	2.66	2.60	3.15	6.25
	Sprinklers	-	-	2.84	2.51	2.63	2.69
16.	Electrical	9.91	9.85	11.31	11.74	13.21	13.98
	SUB TOTAL	$92.44	$104.96	$118.70	$108.72	$117.23	$152.66
	CONTRACTOR'S SERVICES	8.94	10.21	11.47	10.58	11.17	14.85
	Construction Sq.Ft. Cost	$101.38	$115.18	$130.17	$119.30	$128.40	$167.51
	Add for Architect's Fee	7.29	8.34	8.34	8.66	9.11	12.15
	TOTAL Sq.Ft. Cost	$108.67	$123.51	$138.51	$127.96	$137.51	$179.65
	Average Cost per Unit without Architect's Fee	$96,313	$109,419	$117,156	$95,443	$83,458	$108,879
	Average Cost per Unit with Architect's Fee	$103,237	$117,339	$124,659	$102,370	$89,380	$116,776

Add for Porches – Wood - $12.00 SqFt
Add for Interior Garages - $11.40 SqFt

	Average Sq.Ft. Size per Unit	950	950	900	800	650	650
	Average Sq.Ft. Living Space	750	750	750	600	450	450

(1)a Wall Bearing Wood Frame & Wood or Aluminum Façade
(1)b Wall Bearing Masonry & Wood Joists & Brick Façade
(2) Concrete Frame
(3) Post Tensioned Concrete Slabs, Sheer Walls, Window Wall Exteriors, and Drywall Partitions.

Average Square Foot Costs
2011

Commercial

Square Foot Costs do not include Land, Furnishings, and Financing.

Division	Remodeling Office Interior	Office 1-Story (1)	Office 2-Story (2)	Office Up To 5-Story (3)	High Rise (City) Above 5-Story (4)	Parking Ramp (3)
1. General Conditions	$ 3.06	$ 4.96	$ 4.96	7.35	8.80	5.46
2. Site Work & Demolition	2.03	2.90	2.88	2.82	2.64	1.89
3. Concrete	-	8.37	13.00	14.32	9.74	32.03
4. Masonry	-	12.42	11.80	17.82	4.42	1.78
5. Steel	-	4.07	13.48	4.71	3.32	2.56
6. Carpentry and Millwork	9.73	4.50	4.19	6.52	7.36	1.48
7. Moisture Protection	-	8.23	4.11	2.72	3.46	3.83
8. Doors, Windows and Glass	2.04	7.95	7.88	16.03	30.40	1.65
9. Finishes & Sheet Rock	14.11	15.23	14.85	17.15	17.73	2.29
10. Specialties	-	2.02	2.02	2.08	2.08	-
11. Equipment	-	6.45	5.71	4.71	6.42	-
12. Furnishings	-	1.33	1.57	1.74	1.89	-
13. Special Construction	-	-	-	-	-	-
14. Conveying	-	-	3.62	6.24	6.80	3.45
15. Mechanical:						
Plumbing	4.89	7.39	7.93	9.13	9.51	3.20
Heating and Ventilation	2.20	8.29	8.10	10.47	10.53	-
Air Conditioning	2.13	8.24	8.48	10.07	10.04	-
Sprinklers	-	2.64	2.63	2.69	2.71	-
16. Electrical	9.54	11.70	12.89	14.19	16.63	3.69
SUB TOTAL	$49.73	$116.67	$130.11	$150.77	$ 154.48	$63.31
CONTRACTOR'S SERVICES	4.78	9.81	12.60	13.18	13.16	6.08
Construction Sq.Ft. Cost	$54.51	$126.48	$142.71	$163.95	$167.64	$69.39
Add for Architect's Fee	5.40	9.55	9.98	12.04	13.47	5.50
TOTAL Sq.Ft. Costs	$59.92	$136.03	$152.68	$175.99	$181.11	$74.89
Deduct for Tenant Areas Unfinished	-	-	24.74	28.12	32.62	-
Per Car Average	-	-	-	-	-	-$19,683.30
Add: Masonry Enclosed Ramp	-	-	-	-	-	7.54
Add: Heated Ramp	-	-	-	-	-	7.54
Add: Sprinklers for Closed Ramps	-	-	-	-	-	2.92
Deduct for Precast Concrete Ramp	-	-	-	-	-	9.73

(1) Wall Bearing Masonry and Steel Joists and Decks – No Basement
(2) Wall Bearing Masonry and Precast Concrete Decks – No Basement
(3) Concrete Frame
(4) Steel Frame and Window Wall Façade

Average Square Foot Costs
2011

Commercial and Industrial

Square Foot Costs do not include Land, Furnishings, Landscaping, and Financing.
See Division 13 for Metal Buildings

	Division	Store 1-Story (1)	Store 2-Story (2)	Production 1-Story (1)	Warehouse 1-Story (1)	Warehouse and 1-Story Office (1)	Warehouse and 2-Story Office (3)
1.	General Conditions	$ 4.57	$ 4.57	$ 4.40	$ 3.93	$ 4.07	$ 3.70
2.	Site Work & Demolition	2.57	2.01	2.33	2.27	2.69	2.18
3.	Concrete	7.33	14.18	7.45	7.22	8.00	12.26
4.	Masonry	6.18	10.52	8.22	8.24	9.08	5.01
5.	Steel	12.93	6.77	12.93	12.65	13.77	10.39
6.	Carpentry	3.40	3.46	2.21	2.24	3.01	2.61
7.	Moisture Protection	5.79	3.36	6.36	6.13	6.15	3.06
8.	Doors, Windows and Glass	6.70	6.07	5.03	2.72	5.48	4.79
9.	Finishes	7.72	7.26	6.63	2.36	4.13	6.36
10.	Specialties	1.89	1.91	1.46	1.38	1.38	1.38
11.	Equipment	1.82	2.08	1.95	-	1.11	1.25
12.	Furnishings	-	-	-	-	-	-
13.	Special Construction	-	-	-	-	-	-
14.	Conveying	-	4.18	-	-	-	-
15.	Mechanical:						
	Plumbing	5.81	4.99	6.09	3.51	4.42	4.12
	Heating and Ventilation	5.37	5.32	4.13	4.26	5.76	6.63
	Air Conditioning	5.55	5.97	3.93	-	1.47	1.33
	Sprinklers	2.64	2.57	2.49	2.57	2.71	2.63
16.	Electrical	10.89	10.37	11.37	6.91	7.13	6.51
	SUB TOTAL	$91.16	$95.59	$86.98	$66.39	$80.37	$74.21
	CONTRACTOR'S SERVICES	8.77	9.22	8.41	6.42	7.79	7.18
	Construction Sq.Ft. Cost	$99.93	$104.81	$95.39	$72.82	$88.16	$81.39
	Add for Architect's Fees	6.94	7.30	6.67	5.09	6.17	5.69
	TOTAL Sq.Ft. Cost	$106.87	$112.11	$102.06	$77.91	$94.33	$87.08

_____
(1) Wall Bearing Masonry and Steel Joists and Decks – Brick Fronts
(2) Wall Bearing Masonry and Precast Concrete Decks – Brick Fronts
(3) Precast Concrete Wall Panels, Steel Joists, and Deck

Average Square Foot Costs
2011

Educational and Religious

Square Foot Costs do not include Land, Furnishings, Landscaping and Financing.

	Division	Elementary School 1-Story (1)	High School 2-Story (1)	Vocational School 3-Story (1)	College Multi-Story (2)	Gym (3)	Dorm 3-Story (1)	Church (3)
1.	General Conditions	$ 7.91	$ 7.69	$ 7.80	$ 7.69	$ 6.24	$ 6.96	7.46
2.	Site Work	4.70	4.93	4.81	4.70	2.35	3.09	4.54
3.	Concrete	7.45	7.91	5.73	26.70	7.91	5.96	7.22
4.	Masonry	15.48	14.03	15.26	15.92	15.81	14.14	32.18
5.	Steel	13.17	14.86	15.59	5.50	16.13	14.02	2.37
6.	Carpentry	4.83	4.78	4.37	3.20	2.71	4.37	8.85
7.	Moisture Protection	6.64	4.54	5.11	5.13	7.04	6.15	7.32
8.	Doors-Windows-Glass	7.66	7.72	8.06	10.22	7.15	6.64	9.35
9.	Finishes	13.65	11.33	12.72	11.16	9.60	13.07	13.82
10.	Specialties	2.14	1.97	1.97	2.42	1.97	1.88	1.46
11.	Equipment	5.51	5.56	8.18	7.10	10.72	6.76	9.79
12.	Furnishings	-	1.15	1.41	1.46	-	1.41	5.74
13.	Special Construction	-	-	-	-	-	-	-
14.	Conveying	-	-	3.95	3.95	-	3.95	-
15.	Mechanical:							
	Plumbing	12.17	11.75	11.81	15.87	7.39	10.90	7.53
	Heating-Ventilation	15.22	14.93	16.90	15.39	9.72	10.65	10.65
	Air Conditioning	-	-	-	-	-	-	8.10
	Sprinklers	2.69	2.69	2.71	2.74	1.40	2.74	
16.	Electrical	17.73	18.47	19.08	19.39	12.11	13.82	12.13
	SUB TOTAL	$136.94	$134.29	$145.44	$158.54	$118.24	$126.51	148.52
	CONTRACTOR'S SERVICES	13.13	15.05	13.95	15.15	11.36	12.06	14.37
	Constr. Sq.Ft. Cost	$150.07	$149.34	$159.38	$173.69	$129.60	$138.57	162.89
	Add Architect's Fees	10.32	10.14	11.04	12.15	9.00	9.55	11.39
	TOTAL Sq.Ft. Cost	$160.39	$159.48	$170.42	$185.84	$138.61	$148.11	$174.28

(1) Wall Bearing Masonry and Steel Joists and Decks
(2) Concrete Frame and Masonry Façade
(3) Masonry Walls and Wood Roof

Average Square Foot Costs
2011

Medical and Institutional

Square Foot Costs do not include Land, Landscaping, Furnishings, and Financing.

		Clinic 1-Story (1)	Doctors Office 1-Story (1)	Hospital 1-Story (1)	Hospital Multi-Story (2)	Nursing Home 1-Story (1)	Housing for Elderly 1-Story (1)
1.	General Conditions	$ 8.35	$ 8.24	$ 10.97	$ 11.08	$ 9.97	$ 9.24
2.	Site Work	2.89	2.83	4.54	4.58	4.76	4.93
3.	Concrete	8.98	8.92	9.37	28.48	9.74	9.78
4.	Masonry	14.25	12.97	23.61	16.50	16.84	12.36
5.	Steel	11.37	11.24	13.83	3.07	3.08	2.56
6.	Carpentry	8.33	8.21	7.28	4.75	5.13	4.83
7.	Moisture Protection	6.36	6.30	6.84	2.84	6.83	6.79
8.	Doors, Windows and Glass	9.67	10.10	12.86	8.53	11.15	8.91
9.	Finishes	13.78	12.78	17.48	16.88	16.95	14.13
10.	Specialties	1.80	1.80	1.95	1.86	1.86	1.82
11.	Equipment	9.81	9.31	2.32	14.76	7.51	3.33
12.	Furnishings	1.46	1.46	1.55	1.61	1.37	2.91
13.	Special Construction	1.08	1.06	2.16	2.16	1.07	1.08
14.	Conveying	-	-	-	8.02	-	7.35
15.	Mechanical:						
	Plumbing	15.20	16.38	25.01	24.37	16.59	16.05
	Heating and Ventilation	12.50	11.03	16.27	16.78	8.41	6.77
	Air Conditioning	9.60	9.33	16.66	14.49	9.70	3.65
	Sprinklers	2.71	2.69	2.71	2.73	3.10	2.71
16.	Electrical	20.24	16.51	30.70	28.50	18.61	16.39
	SUB TOTAL	$158.40	$151.15	$206.12	$211.99	$152.66	$135.59
	CONTRACTOR'S SERVICES	15.27	14.76	19.77	20.45	14.71	13.11
	Construction Sq.Ft. Cost	$173.67	$165.91	$225.90	$232.44	$167.37	$148.70
	Add for Architect's Fees	12.07	11.69	15.67	16.20	11.65	10.38
	TOTAL Sq.Ft. Cost	$185.74	$177.59	$241.57	$248.64	$179.02	$159.09

Average Cost to Remodel Approximately 50% Cost of New

---

(1) Wall Bearing Masonry and Steel Joists and Decks
(2) Concrete Frame and Masonry Façade

Average Square Foot Costs
2011

Government and Finance

Square Foot Costs do not include Land, Landscaping, Furnishings, and Financing.

	Bank 1-Story (1)	Bank & Office Bldg High-Rise (3)	Govt. Office Bldg Multi-Story (2)	Post Office 1-Story (1)	Court House 3-Story (2)	Jail and Prison (2)
1. General Conditions	$ 9.02	$ 10.36	$ 10.36	$ 8.41	$ 10.25	$ 12.53
2. Site Work	5.56	5.04	4.26	4.24	3.85	5.11
3. Concrete	12.38	29.45	31.86	13.58	27.16	37.13
4. Masonry	17.15	11.97	19.99	11.98	12.83	21.79
5. Steel	6.34	2.84	3.42	4.95	2.96	7.07
6. Carpentry	5.64	5.10	5.24	4.60	3.24	5.80
7. Moisture Protection	4.68	5.26	5.56	4.73	5.85	6.64
8. Doors, Windows and Glass	12.53	29.23	12.80	9.76	9.19	12.03
9. Finishes	18.62	20.81	22.09	18.39	17.92	15.21
10. Specialties	1.91	1.91	3.69	1.99	3.26	7.87
11. Equipment	15.61	10.39	10.28	13.68	13.09	15.27
12. Furnishings	1.73	1.57	-	2.31	-	-
13. Special Construction	-	-	-	-	-	-
14. Conveying	-	3.29	4.12	2.12	3.01	4.07
15. Mechanical						
Plumbing	9.33	10.90	10.82	11.57	14.84	25.74
Heating and Ventilation	7.64	9.47	10.16	7.29	10.84	19.30
Air Conditioning	6.13	6.97	8.84	8.79	15.48	16.05
Sprinklers	2.86	2.74	2.69	2.74	2.74	2.74
16. Electrical	18.83	22.75	23.48	17.98	22.38	34.71
SUB TOTAL	$155.97	$190.06	$189.66	$149.12	$ 178.90	$249.07
CONTRACTOR'S SERVICES	15.23	18.52	18.52	14.53	17.40	24.29
Construction Sq.Ft. Cost	$171.19	$208.58	$208.18	$163.65	$196.30	$273.37
Add for Architect's Fees	12.06	14.67	14.67	11.99	13.78	19.25
TOTAL Sq.Ft. Cost	$183.25	$223.26	$222.86	$175.64	$210.09	$292.61

---

(1) Wall Bearing Masonry and Precast Concrete Decks
(2) Concrete Frame and Masonry Facade
(3) Concrete Frame and Window Wall

UNITED STATES INFLATION RATES
July 1, 1970 to June 30, 2010
AC&E PUBLISHING CO. INDEX

Start July 1

YEAR	AVERAGE %	LABOR %	MATERIAL %	CUMULATIVE %
1970-71	8	8	5	0
1971-72	11	15	9	11
1972-73	7	5	10	18
1973-74	13	9	17	31
1974-75	13	10	16	44
1975-76	8	7	15	52
1976-77	10	5	15	62
1977-78	10	5	13	72
1978-79	7	6	8	79
1979-80	10	5	13	89
1980-81	10	11	9	99
1981-82	10	11	10	109
1982-83	5	9	3	114
1983-84	4	5	3	118
1984-85	3	3	3	121
1985-86	3	2	3	124
1986-87	3	3	3	127
1987-88	3	3	4	130
1988-89	4	3	4	134
1989-90	3	3	3	137
1990-91	3	3	4	140
1991-92	4	4	4	144
1992-93	4	3	5	148
1993-94	4	3	5	152
1994-95	4	3	6	156
1995-96	4	3	6	160
1996-97	3	3	3	163
1997-98	4	3	4	167
1998-99	3	3	2	170
1999-2000	3	2	3	173
2000-2001	4	4	4	177
2001-2002	3	4	3	180
2002-2003	4	5	3	184
2003-2004	3	5	2	187
2004-2005	8	4	10	195
2005-2006	8	6	9	203
2006-2007	8	4	11	212
2007-2008	3	4	3	215
2008-2009	4	3	5	219
2009-2010	3	5	1	222
2010-2011	2	3	2	224